High-level Structures
for Quantum Computing

Synthesis Lectures on Quantum Computing

Editors

Marco Lanzagorta, *U.S. Naval Research Laboratory*
Jeffrey Uhlmann, *University of Missouri-Columbia*

High-level Structures for Quantum Computing
Jarosław Adam Miszczak
2012

Quantum Radar
Marco Lanzagorta
2011

The Complexity of Noise: A Philosophical Outlook on Quantum Error Correction
Amit Hagar
2010

Broadband Quantum Cryptography
Daniel J. Rogers
2010

Quantum Computer Science
Marco Lanzagorta and Jeffrey Uhlmann
2008

Quantum Walks for Computer Scientists
Salvador Elías Venegas-Andraca
2008

© Springer Nature Switzerland AG 2022
Reprint of original edition © Morgan & Claypool 2012

High-level Structures for Quantum Computing

Jarosław Adam Miszczak

ISBN: 978-3-031-01388-1   paperback
ISBN: 978-3-031-02516-7   ebook

DOI 10.1007/978-3-031-02516-7

A Publication in the Springer series
SYNTHESIS LECTURES ON QUANTUM COMPUTING

Lecture #6
Series Editors: Marco Lanzagorta, *U.S. Naval Research Laboratory*
                Jeffrey Uhlmann, *University of Missouri–Columbia*
Series ISSN
Synthesis Lectures on Quantum Computing
Print 1945-9726   Electronic 1945-9734

High-level Structures
for Quantum Computing

Jarosław Adam Miszczak

Institute of Theoretical and Applied Informatics
Polish Academy of Sciences

SYNTHESIS LECTURES ON QUANTUM COMPUTING #6

ABSTRACT

This book is concerned with the models of quantum computation.

Information processing based on the rules of quantum mechanics provides us with new opportunities for developing more efficient algorithms and protocols. However, to harness the power offered by quantum information processing it is essential to control the behavior of quantum mechanical objects in a precise manner. As this seems to be conceptually difficult at the level of quantum states and unitary gates, high-level quantum programming languages have been proposed for this purpose.

The aim of this book is to provide an introduction to abstract models of computation used in quantum information theory. Starting from the abstract models of Turing machine and finite automata, we introduce the models of Boolean circuits and Random Access Machine and use them to present quantum programming techniques and quantum programming languages.

KEYWORDS

quantum computing, models of quantum computation, quantum circuit, quantum finite automaton, quantum Turing machine, quantum RAM, quantum pseudocode, quantum programming language

To my wife Izabela for her support.

Contents

Preface

Classical computers can be programmed using a variety of methods. A user is free to choose the programming language depending on the nature of the problem s/he is working on – if s/he aims for speed and efficiency the natural choice is to use the assembly language or C programming language, if s/he needs to teach his kids s/he is free to choose Logo for this purpose.

The main difference between existing classical programming languages is the distance between the user and the internal architecture of the computing device. The shorter the distance, the better knowledge of the device is required to use it. On the other hand, with the increasing understanding of the computing process, we are able to provide more abstract methods of programming computers.

With the advent of quantum information science, and the discovery of quantum algorithms, the problem of programming quantum computers emerged. Unfortunately, information processing in quantum computers is far from being well-understood. Phenomena like quantum superposition and quantum entanglement, crucial for the computing power of quantum devices, are problematic from the algorithmic point of view. In other words, we still have not learned how to use the power offered by quantum devices.

This book can be divided into two main parts. The first part consists of Chapters 2-5 where we describe theoretical models used in quantum information science. The second part – Chapters 6-9 – is devoted to more pragmatic aspects of using the introduced model for programming quantum computers.

Specifically this book is organized as follows. In Chapter 1 we briefly introduce the areas presented in this book – models of computation, quantum information science, and programming languages. We also sketch how the above areas influence each other. In Chapter 4 we review the model of Turing machine, its variants used to describe nondeterministic and quantum computation, and we give an overview of the relation between the introduced models in terms of computational complexity. In Chapter 4 we introduce the model of quantum circuits, which is commonly used to describe quantum algorithms and communication protocols. In Chapter 3 we introduce quantum finite automata and discuss the relations between classes of quantum languages.

In Chapter 5 we introduce the Quantum Random Access Machine model, which provides a theoretical model used in many quantum programming languages. Chapter 6 describes a layered architecture for translating high-level quantum programming languages into low-level introductions for a physical device used as a quantum machine. In Chapter 7 we introduce the basic requirements for quantum programming languages and we summarize the characteristics of the existing languages. Chapter 8 is devoted to the presentation of two imperative quantum programming languages, while in Chapter 9 we present quantum programming languages based on a functional paradigm. Finally, in Chapter 10 we provide an overview of the possible directions for further development in the area of quantum programming languages and their possible applications.

Jarosław Adam Miszczak
Gliwice, May 2012

Acknowledgments

I would like to thank Jeffrey Uhlmann for encouraging me to write this book and providing valuable remarks. I am also grateful to Michael Morgan for his guidance during the preparation of the text.

Some parts of this book are based on my Ph.D. thesis [1] and previously articles [2, 3, 4, 5]. Chapters 8 and 9 would not be possible without discussions with Piotr Gawron, Bernard Ömer, Wolfgang Mauerer, and Hynek Mlnařík.

Special thanks goes to my wife Izabela. Her support during the preparation of the manuscript was invaluable. She was also helping me with her advice concerning the preliminary version of the text.

Finally, I would like to acknowledge the financial support received during the work on this book from Polish National Science Centre under the grant number N N516 475440.

Jarosław Adam Miszczak
May 2012

CHAPTER 1

Introduction

This book deals with three concepts connecting computer science, theoretical physics, and computer engineering, namely the models of computation, quantum information processing, and programming languages. The aim of this book is to provide a concise introduction to these areas. We aim to show how these concepts, used in classical computer science, are translated into the quantum realm. Moreover, as we are interested in high-level concepts used in quantum information theory, this book aims to introduce these concepts from the quantum programming point of view.

The goal of this book is to provide an introduction to the concepts related to quantum programming. This subject lays in the middle between theoretical aspects of quantum computing and its physical realizations. Readers interested in the first area are advised to consult the books by Nielsen and Chuang [6] and by Lanzagorta and Uhlmann [7], while the reader interested in physical architectures for the realization of quantum information processing should consult comprehensive books by Wiesmann and Milburn [8] and by Metodi, Faruque, and Chong [9].

The purpose of this chapter is to introduce the concepts addressed in the rest of this book. We start by recalling the concept of computability and list the most important models used to study it. Next, we introduce some aspects of quantum information theory. We review possible methods for constructing new algorithms and areas where quantum information processing seems to significantly increase the efficiency. Finally, we introduce basic concepts from the area of programming languages and we review the paradigms used to develop programming languages.

1.1 COMPUTABILITY

Classically, computation can be described using various models. The choice of model depends on the particular purpose or problem. Among the most important models of computation we can point out the following [10]:

- **Turing Machine** introduced in 1936 by Turing and used as the main model in complexity theory [11, 12]. We discuss this model and its quantum generalization in Chapter 2.

- **Boolean circuits** [13, 14] defined in terms of logical gates and used to compute Boolean functions $f : \{0, 1\}^m \mapsto \{0, 1\}^n$; they are used in complexity theory to study circuit complexity. We discuss classical and quantum flavors of this model in Chapter 4.

- **Random Access Machine** [15, 16] which is the example of register machines; this model captures the main features of modern computers and provides a theoretical model for programming languages. This model and its quantum variant are introduced in Chapter 5.

- **Lambda calculus** defined by Church [17] and used as the basis for many functional programming languages [18]. Programming languages based on this model are presented in Chapter 9.

- **Universal programming languages** which are probably the most widely used model of computation [19]. Classical programming languages are briefly described below and their quantum counterparts in Chapters 7, 8, and 9.

It can be shown that all these models are equivalent [11, 12]. In other words the function which is computable using one of these models can be computed using any other model. It is quite surprising since Turing machine is a very simple model, especially when compared with RAM or programming languages.

In particular the model of a multitape Turing machine is regarded as a canonical one. This fact is captured by the Church-Turing hypothesis.

Hypothesis 1.1 Church-Turing   Every function which would be naturally regarded as computable can be computed by a universal Turing machine.

Although stated as a hypothesis this thesis is one of the fundamental axioms of modern computer science. A universal Turing machine is a machine which is able to simulate any other machine. The simplest method for constructing such device is to use the model of a Turing machine with two tapes [11].

The research in quantum information processing is motivated by the extended version of the Church-Turing thesis formulated by Deutsch [20].

Hypothesis 1.2 Church-Turing-Deutsch   Every physical process can be simulated by a universal computing device.

In other words this thesis states that if the laws of physics are used to construct a Turing machine, this model might provide greater computational power when compared with the classical model. Since the basic laws of physics are formulated as quantum mechanics, this improved version of a Turing machine should be governed by the laws of quantum physics.

In this book we review some of these computational models focusing on their quantum counterparts. We start by recalling the basic facts concerning a Turing machine (Chapter 2). This model establishes a clear notion of computational resources like time and space used during computation. It is also used in the formal definitions of other models introduced in this book.

To provide some more insights in the field of quantum complexity we review some information about quantum finite automata (Chapter 3). Although quantum finite automata do not provide a universal model of computation, they can be used to introduce quantum languages and provide a method for comparing the computational power of a large class of quantum and classical devices.

Unfortunately, for practical purposes the notion of Turing machine and finite automata is unwieldy. Even for simple algorithms it requires quite complex description of transition rules. Also, programming languages defined using a Turing machine [21] have a rather limited set of instructions. Thus we discuss more sophisticated methods like Boolean circuits and their quantum counterparts—quantum circuits (Chapter 4). The model of quantum circuits is commonly used to describe quantum algorithms and is tightly connected with the underlying mathematical structure of quantum theory—inputs and outputs are represented by states of quantum (sub)systems and gates are represented by unitary matrices.

The main problem arising in the context of quantum circuits is their limited capability for representing classical parts of the quantum computation—pre- and post-process of results and the preparation of required unitary operations. This problem is addressed by the model of Quantum Random Access Machine (QRAM) (Chapter 5). This model is an extension of the classical Random Access Machine (RAM) model. It introduced the ability to operate on quantum memory and extended the introduction set of the RAM machine by the subset of instructions for operating on this memory.

1.2   QUANTUM INFORMATION THEORY

Quantum information theory is a new, fascinating field of research which aims to use the quantum mechanical description of the system to perform computational tasks. It is based on quantum physics and classical computer science, and its goal is to use the laws of quantum mechanics to develop more powerful algorithms and protocols.

According to the Moore's Law [22] the number of transistors on a given chip is doubled every two years (see Figure 1.1). Since classical computation has its natural limitations in terms of the size of computing devices, it is natural to investigate the behavior of objects in microscale.

Quantum effects cannot be neglected in microscale and thus they must be taken into account when designing future computers. Quantum computation aims not only at taking them into account, but also at developing the methods for controlling them. From this point of view, quantum algorithms and protocols are recipes for how one should control a quantum system to achieve higher efficiency of information processing.

Information processing on quantum computers was first mentioned in 1982 by Feynman [24]. This seminal work was motivated by the fact that the simulation of a quantum system on the classical machine requires exponential resources. Thus, if we could control a physical system at the quantum level we should be able to simulate other quantum systems using such machines.

The first quantum protocol was proposed two years later by Bennett and Brassard [25]. It gave the first example of the new effects which can be obtained by using the rules of quantum theory for information processing. In 1991 Ekert described the protocol [26] showing the usage of quantum entanglement [27] in communication theory.

Today we know that thanks to the quantum nature of photons it is possible to create unconditionally secure communication links [28] or send information with efficiency unachievable while using classical carriers. During the last few years quantum cryptographic protocols have been

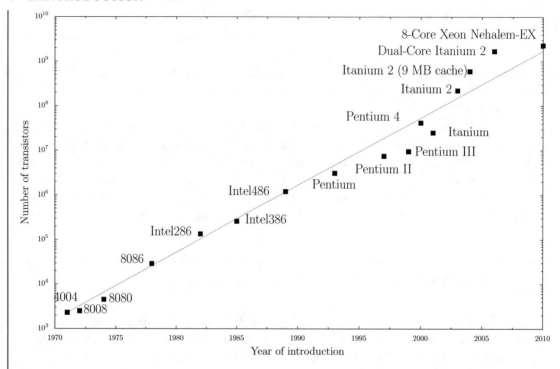

Figure 1.1: Illustration of Moore's hypothesis. The number of transistors which can be put on a single chip grows exponentially. The squares represent microprocessors introduced by Intel Corporation [23]. The dotted line illustrates the rate of growth, with the number of transistors doubling every two years.

implemented in real-world systems. Quantum key distribution is the most promising application of quantum information theory, if one takes practical applications [29, 30, 31] into account.

On the other hand we know that the quantum mechanical laws of nature allow us to improve the solution of some problems [32, 33, 34], and construct random walks [35, 36] and games [37, 38] with new properties.

Quantum walks provide the second promising method for developing new quantum algorithms. They are the counterparts of classical random walks obeying the rules of quantum mechanics [39, 40]. In [41] the quantum algorithm for element distinctness using this method was proposed. It requires $O(n^{2/3})$ queries to determine if the input $\{x_1, \ldots, x_n\}$ consisting of n elements contains two equal numbers. Generalisation of this algorithm, with applications to the problem of subset finding, was described in [42]. Among the other algorithms based on the application of quantum walks one can point out spatial search [43], triangle finding [44], and verifying matrix products [45]. In [46] the survey of quantum algorithms based on quantum walks is presented.

Nevertheless, the most spectacular achievements in quantum information theory up to the present moment are: the quantum algorithm for factoring numbers and the quantum algorithm for calculating discrete logarithms over a finite field proposed in the late nineties by Shor [32, 33]. The quantum algorithm solves the factorization problem in polynomial time, while the best known probabilistic classical algorithm runs in time exponential with respect to the size of input number. Shor's factorization algorithm is one of the strongest arguments for the conjecture that quantum computers can be used to solve in polynomial time problems which cannot be solved classically in reasonable (i.e., polynomial) time.

Taking into account research efforts focused on discovering new quantum algorithms, it is surprising that for the last ten years no similar results have been obtained [47, 48]. One should note that there is no proof that quantum computers can actually solve **NP**-complete problems in polynomial time [49, 50]. This proof could be given by quantum algorithms solving in polynomial time problems known to be **NP**-complete such as k-colorability. The complexity of quantum computation remains poorly understood. We do not have much evidence for how useful quantum computers can be. We know that quantum computers can perform any computation which can be performed on classical computers and that classical computers can reproduce results obtained on quantum computers, i.e., simulate them. However, much remains to be discovered in the area of the relations between quantum complexity classes, such as **BQP**, and classical complexity classes, like **NP**.

One should note, however, that there are many problems for which it is possible to provide a speed-up by using a quantum algorithm. The most comprehensive list with such algorithms is available at [51].

1.3   PROGRAMMING LANGUAGES

We have already mentioned that the main obstacle inhibiting the further development of quantum algorithms and protocols is the counter-intuitive behavior of quantum systems. This behavior seems to be very different from what we observe in the classical world. This problem motivated research for a high-level method which would allow to control the behavior of quantum systems. One of such methods is the development of quantum programming languages.

Programming languages in classical computer science and engineering allow us to overcome the growing complexity of computing systems. They provide a method for communication between the user and the machine [19].

There are thousands of programming languages defined and implemented so far [52]. Some of them are meant to be general-purpose languages, while some are restricted and optimized for a specific area of application. This situation allows us to choose the programming language depending on the particular problem to be solved by the program.

The choice of the programming language used for a particular purpose depends on many factors. Usually the programmer must choose between the speed of writing programs and the speed of running them. Another important factor is a paradigm introduced by a given language, i.e., the set of methods allowing us to reason about programs [53]. For the purpose of this book we are interested in two programming paradigms: imperative and functional.

Most of the existing programming languages follow the imperative style of programming. Among the most important languages from this family we can name Fortran [54], Pascal [55], and C [56]. Programs written in imperative languages are decomposed into steps called commands or instructions. Large programs are divided into procedures or sub-programs to modularize the written code. Programs created in imperative languages reflect, step by step, *how* to solve a given problem. For this reason functional languages are classified as declarative.

Another large group of programming languages support the functional paradigm. Among the functional programming language one can point out Lisp [57], ML [58], and Haskell [59, 60]. In opposition to imperative languages, programs written in functional languages contain the description of *what* shout be computed.

Besides the two families described above, among the popular paradigms one can point out are: logic languages, represented by Prolog [61], and object-oriented languages, represented by Java [62].

Logic languages, along with functional ones, belong to the family of declarative languages. Programs written in functional languages represent the specification of the problem, not the detailed steps required to solve it.

In object-oriented languages programs are collections of objects, which can be accessed using operations defined on them. In some cases the support for the object-oriented paradigm is added as a feature to imperative languages. However, the support for the object-oriented paradigm can be also incorporated into functional languages.

To summarize the above considerations, we can note that many programming languages are described as multi-paradigm. Such languages offer elements commonly associated with one of the paradigms introduced above. As an example of a multi-paradigm language one can point out Python [63], which supports object-oriented, imperative, and functional programming paradigms.

In this book quantum programming languages are introduced with the help of the QRAM model. The discussion of quantum programming languages is presented in Chapters 7, 8, and 9. In these chapters we review basic requirements for quantum programming languages and present two families of such languages. The first one, represented by QCL and LanQ, follows the imperative paradigm. The second one, represented by cQPL and QLM, is based on the principles of functional programming.

CHAPTER 2

Turing machines

The model of a Turing machine is widely used in classical and quantum complexity theory. Despite its simplicity it captures the notion of computability in a universal manner [11, 12]. Thanks to this feature it provides a standard tool for studying the computational complexity of algorithms.

Turing machines can be understood as algorithms for solving problems concerning strings of input elements. They can be used to decide if a string belongs to a language having a given property or they can compute a function by transforming the input string into an appropriate output string.

In this chapter we review a classical model of a Turing machine and we introduce the required notation. We also introduce models of probabilistic and nondeterministic computation, which are crucial in the complexity theory.

Next, we introduce a model of quantum Turing machine. This model provides the first formulation of the quantum computational process. It is also used in the theory of quantum complexity.

2.1 CLASSICAL TURING MACHINE

A Turing machine can operate only using one data structure—the string of symbols. Despite its simplicity, this model can simulate any algorithm with inconsequential loss of efficiency [11].

Let us start by introducing some elementary concepts used in the description of Turing machines. They will be also used in Chapter 3, where we introduce finite-state automata.

In what follows by *alphabet* $A = \{a_1, \ldots, a_n\}$ we mean any finite set of characters or digits. Elements of A are called letters. Set A^k contains all strings of length k composed from elements of A. Elements of A^k are called *words* and the length of the word w is denoted by $|w|$. The set of all words over A is denoted by A^*. Symbol ϵ is used to denote an empty word. The complement of language $L \subset A^*$ is denoted by $\bar{L}$ and it is the language defined as $\bar{L} = A^* - L$.

A classical Turing machine consists of:

- an infinitely long tape containing symbols from the finite alphabet A,

- a head, which is able to read symbols from the tape and write them on the tape,

- memory for storing a program for the machine.

The program for a Turing machine is given in terms of transition function δ. The schematic illustration of a Turing machine is presented in Figure 2.1.

Depending on the definition of the transition function the Turing machine can represent different models of computation: deterministic, probabilistic, or nondeterministic. We start by introducing a deterministic Turing machine and then extend its ability to represent nondeterministic and probabilistic algorithms by altering the form of the transition function.

Formally, the classical deterministic Turing machine is defined as follows [11].

Definition 2.1   Deterministic Turing machine   A deterministic Turing machine M over an alphabet A is a tuple (Q, A, δ, q_0), where

- Q is the set of internal control states,

- $q_0 \in Q$ is an initial state,

- $\delta : Q \times A \mapsto Q \times A \times \{\leftarrow, \downarrow, \rightarrow\}$ is a transition function, i.e., the program of a machine.

It is assumed that the alphabet always contains at least two symbols: $\sqcup$, denoting an empty symbol, and $\triangleleft$, denoting the initial symbol. We also assume that symbols q_a, q_r and h, as well as $\{\leftarrow, \downarrow, \rightarrow\}$, are not elements of $K \cup A$ [1].

By the configuration of machine M we understand a triple $(q_i, x, y), q_i \in Q, x, y \in A^*$. The configuration describes the situation where the machine is in the state q_i, the tape contains the word xy, and the machine starts to scan the word y. If $x = x'$ and $y = b_1 y'$ we can illustrate this situation as in Figure 2.1.

The transition from the configuration c_1 to the configuration c_2 is called a computational step. We write $c \vdash c'$ if δ defines the transition from c to c'. In this case c' is called the successor of c.

A Turing machine can be used to compute the values of functions or to decide about input words. The computation of a machine with input $w \in A^*$ is defined as the sequence of configurations $c_0, c_1, c_2, \ldots$, such that $c_0 = (q_i, \epsilon, w)$ and $c_i \vdash c_{i+1}$. We say that the computation halts if some c_i has no successor or for configuration c_i, the state of the machine is q_a (machine accepts input) or q_r (machine rejects input).

The computational power of the Turing machine has its limits. Let us define two important classes of languages.

Definition 2.2   A set of words $L \in A^*$ is a recursively enumerable language if there exists a Turing machine accepting input w iff $w \in L$.

Definition 2.3   A set of words $L \in A^*$ is a recursive language if there exists a Turing machine M such that

- M accepts w iff $w \in L$,

- M halts for any input.

---

[1]Note that this is a matter of convention (see e.g. [64]).

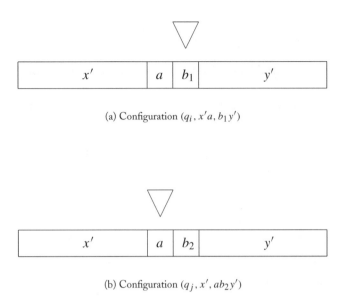

(a) Configuration $(q_i, x'a, b_1 y')$

(b) Configuration $(q_j, x', ab_2 y')$

Figure 2.1: Computational step of the Turing machine. Configuration $(q_i, x'a, b_1 y')$ is presented in (a). If the transition function is defined such that $\delta(q_i, b_1) = (q_2, b_2, -1)$ this computational step leads to configuration $(q_j, x', ab_2 y')$ (see (b)).

The computational power of the Turing machine is limited by the following theorem.

Theorem 2.4 *There exists a language H which is recursively enumerable but not recursive.*

Language H used in the above theorem is defined in halting problem [11]. It consists of all words composed of words encoding Turing machines and input words for these machines, such that a particular machine halts on a given word. A universal Turing machine can simulate any Turing machine. Therefore, for a given input word encoding the machine and an input for this machine, we can easily perform the required computation.

A deterministic Turing machine is used to measure time complexity of algorithms. Note that if for some language there exists a Turing machine accepting it, we can use this machine as an algorithm for solving this problem. Thus we can measure the running time of the algorithm by counting the number of computational steps required for a Turing machine to output the result.

The time complexity of algorithms can be described using the following definition.

Definition 2.5   Complexity class **TIME**$(f(n))$ consists of all languages L such that there exists a deterministic Turing machine running in time $f(n)$ accepting input w iff $w \in L$.

In particular, complexity class **P** defined as

$$\mathbf{P} = \bigcup_k \mathbf{TIME}(n^k),\tag{2.1}$$

captures the intuitive class of problems which can be solved *easily* on a Turing machine.

2.2   NONDETERMINISTIC AND PROBABILISTIC COMPUTATION

Since one of the main features of quantum computers is their ability to operate on the superposition of states, we can easily extend the classical model of a probabilistic Turing machine and use it to describe quantum computation. Since in general many results in the area of algorithm complexity are stated in the terms of a nondeterministic Turing machine we start by introducing this model.

Definition 2.6   Nondeterministic Turing machine   A nondeterministic Turing machine M over an alphabet A is a tuple (Q, A, δ, q_0), where

- Q is the set of internal control states,

- $q_0 \in Q$ is the initial state,

- $\delta \subset Q \times A \times Q \times A \times \{\leftarrow, \downarrow, \rightarrow\}$ is a relation.

As previously, we assume that the special symbols used in the above definition are not elements of $Q \cup A$.

One should note that in the above definition the function (see Definition 2.1) was replaced by a relation. In other words, in one step of computation the machine can move from one state to more than one state and thus the computational path of the machine can branch.

The last condition in the definition of a nondeterministic machine is the reason for its power. It also requires changing the definition of acceptance by the machine.

We say that a nondeterministic Turing machine accepts input w if, for some initial configuration (q_i, ϵ, w), the computation leads to configuration (q_a, a_1, a_2) for some words a_1 and a_2. Thus a nondeterministic machine accepts the input if there exists some computational path defined by transition relation δ leading to an accepting state q_a.

The model of a nondeterministic Turing machine is used to define complexity classes **NTIME**.

Definition 2.7   Complexity class **NTIME**$(f(n))$ consists of all languages L such that there exists a nondeterministic Turing machine running in time $f(n)$ accepting input w iff $w \in L$. The most

prominent example of these complexity classes is **NP**, which is the union of all classes **NTIME**(n^k), i.e.,

$$\mathbf{NP} = \bigcup_k \mathbf{NTIME}(n^k).\tag{2.2}$$

A nondeterministic Turing machine is used as a theoretical model in complexity theory. However, it is hard to imagine how such a device operates. One can illustrate the computational path of a nondeterministic machine as shown in Figure 2.2 [11].

Figure 2.2: Schematic illustration of the computational paths of a nondeterministic Turing machine [11]. Each circle represents the configuration of the machine. The machine can be in many configurations simultaneously.

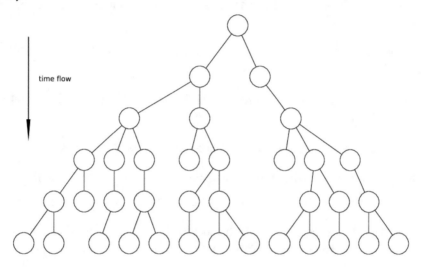

Since our aim is to provide the model of a physical device, we restrict ourselves to a more realistic model. We can do that by assigning a number representing probability to each element of the relation. In this case we obtain the model of a probabilistic Turing machine.

Definition 2.8 Probabilistic Turing machine   A probabilistic Turing machine M over an alphabet A is a tuple (Q, A, δ, q_0), where

- Q is the set of internal control states,

- $q_0 \in Q$ is the initial state,

- $\delta : Q \times A \times Q \times A \times \{\leftarrow, \downarrow, \rightarrow\} \mapsto [0, 1]$ is a transition probability function i.e.,

$$\sum_{(q_2, a_2, d) \in Q \times A \times \{\leftarrow, \downarrow, \rightarrow\}} \delta(q_1, a_1, q_2, a_2, d) = 1. \tag{2.3}$$

For a moment we can assume that the probabilities of transition used by a probabilistic Turing machine can be represented only by rational numbers. We do this to avoid the problems with machines operating on arbitrary real numbers. We will address this problem when extending the above definition to the quantum case.

We have already pointed out that Turing machines are equivalent to algorithms—the construction of a Turing machine for solving a given problem is equivalent to finding an algorithm solving this problem. Now, using the number of steps executed by this machine, one can describe the time complexity of the problem.

The time complexity of computation can be measured in terms of the number of computational steps of the Turing machine required to execute a program. Among important complexity classes we have chosen to point out:

- **P** – the class of languages for which there exists a deterministic Turing machine running in polynomial time,

- **NP** – the class of languages for which there exists a nondeterministic Turing machine running in polynomial time,

- **PP** – the class of decision problems solvable by an **NP** machine such that:

    - if the answer is '*yes*' then at least 1/2 of computation paths accept the input,

    - if the answer is '*no*' then less than 1/2 of computation paths accept the input.

- **RP** – the class of languages L for which there exists a probabilistic Turing machine M such that: M accepts input w with probability at least $\frac{1}{2}$ if $w \in L$ and always rejects w if $w \notin L$,

- **coRP** – the class of languages L for which $\bar{L}$ is in **RP**,

- **ZPP** – **RP** $\cap$ **coRP**.

More examples of interesting complexity classes and computational problems related to them can be found in [65].

2.3   QUANTUM TURING MACHINE

A quantum Turing machine was introduced by Deutsch in [20]. This model is equivalent to a quantum circuit model [66, 67]. However, it is very inconvenient for describing quantum algorithms since the state of a head and the state of a tape are described by state vectors.

A quantum Turing machine consists of:

- processor: M 2-state observables $\left\{ n_i | i \in_{\mathbb{Z}_M} \right\}$,

- memory: infinite sequence of 2-state observables $\{ m_i | i \in \mathbb{Z} \}$,

- observable x, which represents the address of the current head position.

The state of the machine is described by the vector $|\psi(t)\rangle = |x; n_0, n_1, \ldots; m\rangle$ in the Hilbert space $\mathcal{H}$ associated with the machine.

At the moment $t = 0$ the state of the machine is described by the vectors $|\psi(0)\rangle = \sum_m a_m |0; 0, \ldots, 0; \ldots, 0, 0, 0, \ldots\rangle$ such that

$$\sum_i |a_i|^2 = 1. \tag{2.4}$$

The evolution of the quantum Turing machine is described by the unitary operator U acting on $\mathcal{H}$.

A classical probabilistic (or nondeterministic) Turing machine can be described as a quantum Turing machine such that, at each step of its evolution, the state of the machine is represented by the base vector.

The formal definition of the quantum Turing machine was introduced in [49].

It is common to use real numbers as amplitudes when describing the state of quantum systems during quantum computation. To avoid problems with an arbitrary real number we introduce the class of numbers which can be used as amplitudes for amplitude transition functions of the quantum Turing machine.

Let us denote by $\widetilde{\mathbb{C}}$ the set of complex numbers $c \in \mathbb{C}$, such that there exists a deterministic Turing machine, which allows to calculate $\mathrm{Re}\,(c)$ and $\mathrm{Im}\,(c)$ with accuracy $\frac{1}{2^n}$ in time polynomial in n.

Definition 2.9 Quantum Turing Machine A quantum Turing machine (QTM) M over an alphabet A is a tuple (Q, A, δ, q_0), where

- Q is the set of internal control states,

- $q_0 \in Q$ is the initial state,

- $\delta : Q \times A \times Q \times A \times \{\leftarrow, \downarrow, \rightarrow\} \mapsto \widetilde{\mathbb{C}}$ is a transition amplitude function i.e.,

$$\sum_{(q_2, a_2, d) \in Q \times A \times \{\leftarrow, \downarrow, \rightarrow\}} |\delta(q_1, a_1, q_2, a_2, d)|^2 = 1. \tag{2.5}$$

Reversible classical Turing machines (i.e., Turing machines with reversible transition function) can be viewed as particular examples of quantum machines. Since any classical algorithm can be transformed into reversible form, it is possible to simulate a classical Turing machine using a quantum Turing machine.

2.4   MODIFICATIONS OF THE BASE MODEL

Since the introduction of the quantum Turing machine, many modifications of the base model appeared. The research in this area is motivated by several factors. First, new methods of quantum computation were introduced, e.g., measurement-based quantum computation, where there is a clear need for representing classical and quantum data using a single theoretical model. The second reason stems from the need for representing quantum computation involving, besides unitary transform, also operations executed on arbitrary (mixed) quantum states.

2.4.1   GENERALIZED QUANTUM TURING MACHINE

The possible transformations allowed in quantum mechanics are described by completely positive transformations on the set of mixed quantum states (statistical mixtures of pure states represented by ket vectors) [6]. While usually quantum computation is described in the terms of pure states, the need for the introduction of mixed states is motivated by the irreversible transformation occurring during the interactions of quantum systems with the environment.

Taking the above into account, the major drawback appearing in the definition of a quantum Turing machine (Definition 2.9) is the restriction of the set of available operations to the set of unitary (reversible) transformations.

This restriction can be overcome by allowing a more general form of quantum evolution and by allowing the internal states and symbols appearing on the input state to be represented by density operators.

Such a model of computation was introduced in [68] as generalized quantum Turing machine. This model is defined as follows.

Definition 2.10   Generalized Quantum Turing Machine   Generalized quantum Turing machine is a tuple $(Q, A, \mathcal{H}, \Delta)$, where

- Q is the set of internal states represented by density operators,

- $q_0 \in Q$ is the initial state,

- Δ is a quantum channel i.e., map on the state of mixed states (density operators).

The quantum channel Δ in the above definition represents a program of the machine and it acts on the compound space consisting of the state of the tape, the initial state, and the state of the reading head. If we denote the underlying spaces by $\mathcal{H}_A$, $\mathcal{H}_Q$, and $\mathcal{H}_Z$, respectively, then the map Δ is given by a completely positive operation on

$$\mathcal{H}_Q \otimes \mathcal{H}_A \otimes \mathcal{H}_Z.$$

2.4.2   CLASSICALLY CONTROLLED QUANTUM TURING MACHINE

The models of computation discussed previously have one common drawback—they do not allow incorporating classical control structures required to perform any quantum computation. As we will see in Chapter 5 this problem is addressed by the Quantum Random Access Machine (QRAM) model. It is, however, possible to introduce a modification of quantum Turing machine which allows to incorporate classical control structures.

Classically controlled quantum Turing machine (CQTM), introduced in [69], consists of a quantum tape and classical internal sets. A quantum tape is used to operate on quantum data. Classical states are used to formalize the classical control required in any practical model of quantum computation.

Formally, the model of CQTM is introduced as follows.

Definition 2.11   Classically controlled Quantum Turing Machine   Classically controlled quantum Turing machine is a quintuple $(K, A_C, A_Q, U, \delta, k_0, k_a, k_r)$, where

- K is a finite set of classical states,

- $k_0, k_a, k_r \in K$ are initial, accepting and rejecting states,

- A_C is a finite alphabet of classical outcomes,

- A_Q is a finite alphabet of quantum basis states,

- U is a finite set of quantum transformations, consisting of elements of the form $U_i = \{M_k : k \in A_C\}$ such that

$$\sum_{k \in A_C} M_k^\dagger M_k = \mathbb{I}. \tag{2.6}$$

- δ is a classical transition function

$$\delta : K \times A_C \mapsto (K \cup) \tag{2.7}$$

In the above definition, the set of admissible transformations is defined using the Kraus form of quantum channels [6]. One should note that the operators M_k in this definition can map $\mathbb{C}^{d^n}$ onto $\mathbb{C}^{d^m}$ if one allows operations on d-dimensional systems (qudits) and transform n-qudit registers into m-qudit registers.

The use of quantum channels allows to incorporate any physically allowed operation. Thanks to this fact, CQTM can operate on the measurement results obtained during the computational process.

2.5   QUANTUM COMPLEXITY

Quantum Turing machine allows for rigorous analysis of algorithms. This is important since the main goal of quantum information theory is to provide some gain in terms of speed or memory with respect to classical algorithms. It should be stressed that, at the moment, no formal proof has been given that a quantum Turing machine is more powerful than a classical Turing machine [50].

In this section we give some results concerning quantum complexity theory. See also [49, 70] for an introduction to this subject.

In analogy to the classical case it is possible to define complexity classes for the quantum Turing machine. The most important complexity class in this case is **BQP**.

Definition 2.12   Complexity class **BQP** contains languages L for which there exists a quantum Turing machine running in polynomial time such that, for any input word x, this word is accepted with probability at least $\frac{3}{4}$ if $x \in L$ and is rejected with probability at least $\frac{3}{4}$ if $x \notin L$.

Class **BQP** is a quantum counterpart of the classical class **BPP**.

Definition 2.13   Complexity class **BPP** contains languages L for which there exists a nondeterministic Turing machine running in polynomial time such that, for any input word x, this word is accepted with probability at least $\frac{3}{4}$ if $x \in L$ and is rejected with probability at least $\frac{3}{4}$ if $x \notin L$.

Since many results in complexity theory are stated in terms of oracles, we define an oracle as follows.

Definition 2.14   An oracle or a black box is an imaginary machine which can decide certain problems in a single operation.

We use notation $\mathbf{A^B}$ to describe the class of problems solvable by an algorithm in class **A** with an oracle for the language **B** [71].

It was shown in [49] that the quantum complexity classes are related as follows.

Theorem 2.15   *Complexity classes fulfill the following inequality*

$$\mathbf{BPP} \subseteq \mathbf{BQP} \subseteq \mathbf{P^{\#P}}. \tag{2.8}$$

Complexity class **#P** consists of problems of the form *compute f(x), where f is the number of accepting paths of an* **NP** *machine.* For example problem **#SAT** formulated below is in **#P**.

Problem 2.16   #SAT   For a given Boolean formula, compute how many satisfying true assignments it has.

Complexity class $\mathbf{P^{\#P}}$ consists of all problems solvable by a machine running in polynomial time which can use an oracle for solving problems in $\mathbf{\#P}$.

Complexity ZOO [65] contains the description of complexity classes and many famous problems from complexity theory. The complete introduction to complexity theory can be found in [11]. Theory of **NP**-completeness with many examples of problems from this class is presented in [72].

Many important results and basic definitions concerning quantum complexity theory can be found in [49]. The proof of equivalence between quantum circuit and quantum Turing machine was given in [66]. An interesting discussion of quantum complexity classes and relation of **BQP** class to classical classes can be found in [50].

2.6   FANTASY QUANTUM COMPUTING

We have already mentioned that the model of a Turing machine is very useful for studying the complexity of the computational process. For this reason it is sometimes desirable to introduce non-realistic models of computations and explore their computational power. One such model is a quantum Turing machine with post-selection (PostQTM) [73]. This machine has the ability to *postselect*, i.e., reject the outcomes of computation which result in a given event does not occur.

This construction shows that quantum computing can be also useful for studying the properties of classical complexity classes. The complexity class of problems solvable efficiently by PostQTM is **PostBQP** defined by Aaronson in [73].

Definition 2.17   PostBQP   A language L is said to be in **PostBQP**, if there exists a uniform size of polynomial quantum circuits $\{C_n\}_N \geq 1$ such that for all inputs x

- after the application of C_n to $|0 \ldots 0\rangle \otimes |x\rangle$, the first qubit has a non-zero amplitude at $|1\rangle$,

- if $x \in L$ then the probability of measuring $|1\rangle$ at the first qubit and then $|1\rangle$ at the second qubit is at least $2/3$,

- if $x \notin L$ then the probability of measuring $|1\rangle$ at the first qubit and then $|1\rangle$ at the second qubit is at least $1/3$.

The condition in the above definition means that one rejects the outcomes of computation which result in a state where a given event does not occur, i.e., postselects results.

One should note that **PostBQP** is simply a variant of **BQP** extended with the ability to postselect results.

The most important fact about this complexity class is given by the following theorem.

Theorem 2.18   PostBQP = PP

One of the reasons for introducing such a model of computation comes from the considerations related to the fundamentals of quantum mechanics. PostQTM can be simulated if we allow for the linear, but not unitary evolution, of quantum systems. As PostQTM can solve efficiently the problems which are known to be classically difficult (i.e., problems in **PP**), this may explain why the evolution of quantum systems is unitary.

2.7   SUMMARY

The model of a quantum Turing machine is crucial as a theoretical background for the development of other computational models used in quantum computation, e.g., the quantum circuit model described in Chapter 4. However, it is not very useful for developing new quantum algorithms and protocols.

The main obstacle in using this model for programming quantum computers is the lack of data types which can be used to represent the input and the result of a computation. A quantum Turing machine operates on a single type of data, namely a list of symbols.

Also the programming of a Turing machine, classical or quantum, is a very cumbersome task. In order to show that a given function (e.g., addition or multiplication) can be computed by a Turing machine, one has to write a program for the machine that calculates this function. This is feasible, however, only for very simple functions. In most cases to show that a function can be computed by a Turing machine, it is reasonable to show that it can be computed by a model which is equivalent to the Turing machine [74].

The study of quantum complexity theory is the most important application of the quantum Turing machine. At the moment of writing the exact relation between classical and quantum complexity classes is still unknown. The question if the quantum computer can indeed be used to efficiently solve problems untraceable by classical computers is of great importance. For this reason we return to this topic once again in Chapter 3, where we introduce quantum finite automata and compare them with classical finite automata.

CHAPTER 3

Quantum Finite State Automata

In this chapter we define quantum finite automata—another model of computation used to study quantum computation. Again, as in the case of the Turing machine, we start with the classical version of this model, namely deterministic finite automaton, and subsequently upgrade it. Firstly, by considering nondeterminism and stochastic transitions, and then, by allowing for the transition to be described by unitary matrices.

Quantum finite state automata (QFA) provide an appealing model of quantum computation. As the name suggests, this model of computation is based on the finite state automata model used widely in classical theory of computation [64]. To be more precise, quantum finite automata provide a quantum analogue of probabilistic finite automata. As such quantum finite automata generalize the concept of Markov chain.

They can be used to process strings of input symbols by the devices which are able to operate on data encoded in quantum carriers. This allows us to introduce a concept of a quantum language and provides the means for comparing the complexity of processing different classes of languages in classical and quantum cases. As the understanding of the actual power of quantum devices is one of the most exciting problems in the theory of quantum information, the quantum automata are an important model used for understanding the computing capabilities offered by quantum computers.

We start this chapter by introducing the concept of classical finite state automata and we show how it can be generalized to the quantum domain. As in the case of the Turing machine, the quantum version of finite automaton is obtained by applying quantum rules of calculating probabilities to the probabilistic version of the classical model.

3.1 FINITE AUTOMATA

Finite automata or finite-state machines [64, 75, 76] provide one of the simplest models of computation. Devices of this type read a string of symbols as an input. The only type of output they are able to deliver is the answer if the input string is considered acceptable. The devices of this type are called acceptors.

Even if such a simple model seems to be of little use, finite automata are used in many areas of engineering. The first paper on finite automata, written by Kleen in 1956 [77], concerned the study of nerve nets. Today among the most interesting applications of finite automata is lexical analysis and pattern matching, used in the design and implementation of programming languages.

Below we focus on the models describing acceptor or recognizer devices. One should note, however, that in some situations it is desirable to consider transducer devices, i.e., finite state machines that are able to generate output based on input string (Moore model) [76] or input string and a state (Mealy model) [75]. Such devices are used in the area of computational linguistics or in control applications.

3.1.1　DETERMINISTIC FINITE AUTOMATA

We start by introducing the model of deterministic finite automata, which provides basic ingredients used in the formal definitions of other models described in this chapter. This model can also be used to introduce the notion of regular languages.

Definition 3.1　Deterministic finite automaton (DFA)　A deterministic finite automaton M is a tuple $M = (Q, \Sigma, \delta, s_0, F)$ where

- Q is a finite set of states,

- Σ is an alphabet of input symbols,

- $\Delta : Q \times \Sigma \mapsto Q$ is a transition function,

- $s_0 \in Q$ is the initial or starting state,

- $F \subseteq Q$ is the set of final states.

Alphabet Σ consists of all possible input symbols, which can appear on the input tape. The transition function describes the rules according to which the automaton changes its state after reading the subsequent symbols from the input tape.

Figure 3.1 illustrates the computation process of deterministic finite state automaton. A reading head is initially placed over the leftmost symbol on the input tape. In each step the machine scans one symbol from the input tape and changes its state according to the current state and the scanned symbol. After this the head moves one step to the right.

The configuration of the automaton M is described by its current state and the portion of the input string which is not yet processed by the machine. Thus, a configuration is an element of $Q \times \Sigma^*$. For example, the automaton in Figure 3.1 is in the configuration $(q_1, abcabc)$.

A string $\omega \in \Sigma^*$ is said to be accepted by M if and only if there exists a state $f \in F$ such that after some finite number of steps an automaton starting in configuration (s, ω) yields (f, ϵ). A language consisting of all strings accepted by M is denoted by $L(M)$.

Definition 3.2　Equivalence of automata　We say that two automata are equivalent if and only if $L(M_1) = L(M_2)$.

In the definition of deterministic finite-state automaton (Definition 3.1) we assumed that the device can only move the reading head to the right, i.e., it is unable to review the portion of the input string which has already been processed. The model of deterministic finite-state machine with the

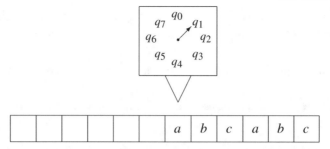

Figure 3.1: Configuration of the deterministic finite state automata. The automaton can be visualised as the device consisting of the head, used to read input symbols, and the memory for storing the internals states. These states are altered upon the readout of a symbol in a manner described by the transition function. In the above picture, the machine is in the configuration $(q_1, abcabc)$, which means that its current state is q_1 and the input tape contains string $abcabc$.

ability to move the reading head in both directions is known as two-way deterministic finite-state automaton.

The computational power offered by 1-way and 2-way classical finite automata is identical. We will see, however, that this is not the case when one extends these models according to quantum rules. That is why the 2-way finite automata are important in the quantum context.

Definition 3.3 Two-way deterministic finite automaton (2DFA)   A deterministic finite automaton M is a tuple $M = (Q, \Sigma, \delta, s_0, F)$ where

- Q is a finite set of states,

- Σ is an alphabet of input symbols,

- $\Delta : Q \times \Sigma \mapsto Q \times \{\leftarrow, \downarrow, \rightarrow\}$ is a transition function,

- $s_0 \in Q$ is the initial or starting state,

- $F \subseteq Q$ is the set of final states.

The computational process of the 2DFA is very similar to the computation of DFA. The only difference is that after reading the input symbol and changing the internal state, the 2DFA is able to change the position of the reading head by moving it to the left ($\leftarrow$), to the right ($\rightarrow$) or not moving it at all ($\downarrow$).

Taking into account the above definition, it is easy to see that the class of two-way deterministic quantum automata is a superset of the set of (one-way) deterministic finite automata. Moreover, the following theorem states that both classes of automata have the same computational power.

Theorem 3.4 Rabin and Scott [78], Shepherdson [79]   *For any two-way deterministic finite automaton there exists an equivalent one-way deterministic finite automaton.*

The above theorem suggests that the class of languages recognized by finite state automata is in some sense special. The languages recognized by deterministic finite state automata are called regular languages.

This class of languages is important for comparing the computational power of classical and quantum automata.

In order to define regular languages we introduce the notion of regular expressions.

Definition 3.5   Regular expression   A regular expression over the alphabet Σ is a string over the alphabet $\Sigma \cup \{\emptyset, |, {}^*, (,)\}$, which can be obtained from the following rules

- $\emptyset$ and any element of Σ is a regular expression,

- if α and β are regular expressions, then so is $(\alpha\beta)$,

- if α and β are regular expressions, then so is $(\alpha|\beta)$,

- if α is a regular expressions, then so is (α^*),

where operations $|$, * and juxtaposition are interpreted as

- * – Kleen star,

- $|$ – alternation of two sets,

- $(\alpha\beta)$ – concatenation of two strings.

One should note that the operations of Kleen star, alternation and concatenation used in the above definition, are normally defined for languages (i.e., sets). Formally, the relation between regular expressions and languages is defined by the function $\mathcal{L}$ mapping regular expressions to languages, such that

- $\mathcal{L}(\emptyset) = \emptyset$ and $\mathcal{L}(a) = \{a\}$,

- $\mathcal{L}(\alpha^*) = \mathcal{L}(\alpha)^*$,

- $\mathcal{L}(\alpha|\beta) = \mathcal{L}(\alpha)|\mathcal{L}(\beta)$,

- $\mathcal{L}(\alpha\beta) = \mathcal{L}(\alpha)\mathcal{L}(\beta)$.

Using the above definition we introduce a class of regular languages as follows.

Definition 3.6   Regular language   A class of regular languages over an alphabet Σ consists of all languages L such that $L = \mathcal{L}(\alpha)$ for some regular expression $\alpha \in (\Sigma \cup \{\emptyset, |, {}^*, (,)\})^*$

The class of regular languages is exactly the class of languages recognized by deterministic finite automata.

Theorem 3.7   *Language L is called regular if and only if there exists a deterministic finite automaton that recognizes L.*

Based on the above, one can see that the deterministic (one-way and two-way) finite automata are equivalent to regular expressions.

3.1.2   NONDETERMINISTIC FINITE AUTOMATA

The natural generalization of the deterministic finite state automata is provided by the nondeterministic finite state automata. Unlike the case of Turing machines, this model does not posses a greater computational power than its deterministic counterpart. However, it can be helpful in introducing the probabilistic version of finite automata.

The model of nondeterministic finite automata was first introduced by Rabin and Scott [78].

Definition 3.8   Nondeterministic finite automaton   A deterministic finite automaton M is a tuple $M = (Q, \Sigma, \Delta, s_0, F)$ where

- Q is a finite set of states,

- Σ is an alphabet of input symbols,

- $\Delta \subset Q \times \Sigma \times Q$ is a transition relation,

- $s_0 \in Q$ is the initial or starting state,

- $F \subseteq Q$ is the set of final states.

Alternatively, in the above definition one can use a transition function of the form $\Delta : Q \times \Sigma \to 2^Q$, where 2^Q denotes the power set for Q. This method of defining allows us to stress that there can exist many computational paths leading from a given configuration. Thus there are many allowed transitions for a given state and a given input symbol (compare Figure 2.2 in Chapter 2).

The transition from deterministic to non-deterministic automata is very similar to the case of the Turing machine. In the case of automata, however, the nondeterminism does not increase the computational power of the machine.

This fact, expressed by the following theorem, was proved in [78], but it can be found in many textbooks on the theory of computation (see e.g., [80]).

Theorem 3.9   Rabin and Scott [78]   *For each nondeterministic finite automaton, there exists an equivalent deterministic finite automaton.*

The above theorem states that both classes of automata recognize exactly the same class of languages, namely regular languages. Thus, we can use the following definition of a regular language, which is equivalent to the Definition 3.6.

Definition 3.10   Regular language   A language L is called regular if and only if there exists a finite state automaton, deterministic or non-deterministic, that recognizes L.

Regular languages are commonly used in pattern matching and many modern, general purpose programming languages provide a function for operating on regular expressions [81].

The structure of words generated by regular expressions is described by a pumping lemma for regular languages [82].

Lemma 3.11   Pumping lemma for regular languages   *Let L be a regular language. Then there exists a constant $p \geq 1$, such that for any $\omega \in L$ the following condition*

$$|\omega| \geq p \Rightarrow \left[\exists_{u,v,w \in L}(\omega = uvw \wedge |uw| \leq p \wedge |v| \geq 1) \forall_{i \geq 0} uv^i w \in L \right]. \qquad (3.1)$$

holds. Here by $|\omega|$ we denote the length of word ω.

Informally, the pumping lemma states that if a regular language L contains words of a length greater than some constant p, then each such word can be expressed by repeating some finite sequence of letters.

The variant of this lemma can also be stated for quantum counterparts of regular languages. In quantum as well as in classical automata theory, the pumping lemma allows us to show that a given language is not in a given class.

The transition between deterministic and nondeterministic automata can be illustrated with the use of the matrix representation. Transitions between states in a finite-state automaton can be described in the matrix form. More specifically, to each letter in the input alphabet one can assign a $n \times n$, where n is the number of states.

In the case of deterministic automata, the matrix associated with input symbols contains only one element in each column and this element is equal to 1.

In the case of nondeterministic automata, the matrices can contain more than one element in each column and these elements, describing a number of computational paths, can be greater than 1.

3.1.3   PROBABILISTIC AUTOMATA

The concept of quantum finite state automaton can be introduced as a generalization of a probabilistic finite-state automaton.

The model of probabilistic automaton was introduced by Rabin [83]. This model provides a natural generalization of the concept of nondeterministic automaton.

Definition 3.12   Probabilistic finite state automaton (PFA)   A probabilistic automaton M is a tuple $M = (Q, \Sigma, \Delta, s_0, F)$ where

- Q is a finite set of states,

- Σ is an alphabet of input symbols,

- $\Delta : Q \times \Sigma \mapsto P(Q)$ is a transition function, where $P(Q)$ is a probability vector indexed by the elements of Q.

- $s_0 \in P(Q)$ is the initial or starting state,

- $F \subseteq Q$ is the set of final states.

In the above definition the transition function associates with each pair $(q_i, s_i) \in Q \times \Sigma$ a vector of probabilities (i.e., non-negative numbers summing to 1). This vector describes the possible transition allowed after the occurrence of (q_i, s_i).

One can also note that the initial state of the automaton is described by a probability vector. In such situation, the automaton starts in the state described by a mixture of elements of Q.

Probabilistic automata are sometimes called *Rabin automata*. This name is, however, usually reserved for a special type of probabilistic automata.

Definition 3.13 Rabin automaton   The Rabin automaton is a probabilistic finite state automaton such with an initial state of $(q, 0, \ldots, 0)$.

Probabilistic automata can recognize a larger class of languages than deterministic finite automata. The languages recognized by probabilistic automata are called *stochastic languages*.

Definition 3.14 Stochastic languages   A language L is called stochastic, if and only if there exists a probabilistic finite automaton that recognizes L.

In the matrix representation, probabilistic automata are described by a collection of stochastic matrices. More precisely, matrices are left stochastic, which means that each of the columns consists of non-negative real numbers summing to 1. Such matrices are used to describe Markov chains.

3.2   QUANTUM FINITE AUTOMATON

Quantum mechanics is a probabilistic theory and it can only provide information about the probabilities of results obtained in the experiments. For this reason it is natural to ask if the model of finite automata can gain more computational power by enriching it with the elements known from quantum mechanics, namely the ability to operate on superpositions of states and the requirement that the time evolution of the state has to be reversible.

The starting point for such extension is the model of probabilistic finite automata. Any type of finite state automata can be characterized by a characteristic function $\chi_L : \Sigma^\star \mapsto \{0, 1\}$ defined as

$$\chi_L(\omega) = \begin{cases} 1 & \omega \in \Sigma^\star \\ 0 & \omega \notin \Sigma^\star \end{cases} . \tag{3.2}$$

Any language can be identified with its characteristic function. In this approach one can define a language by using a mapping from words to probabilities

$$\chi_L^Q(\omega) \mapsto [0, 1]. \tag{3.3}$$

As in the case of classical automata, quantum automata can be divided into two groups: one-way quantum automata (1QFA) and two-way quantum automata (2QFA). However, unlike in the classical case, there is a significant difference in the computational power offered by these two models.

Among one-way quantum automata one can distinguish two basic types: measurement-many (MM-1QFA) [84] and measurement-once (MO-1QFA) [85]. The most important class of two-way quantum automata is the class of measure-many quantum finite automata (MM-2QFA) [84].

The transition from probabilistic to quantum automata can be also described using the matrix representation. In the case of quantum automata, matrices describing an automaton correspond to the transition between quantum states. As such they must be unitary.

3.2.1   MEASURE-ONCE QUANTUM FINITE AUTOMATON

The simplest model of quantum automata was introduced by Moore and Crutchfield [85]. In this model the transitions between states are given in the terms of unitary operators. The state of the device is measured after processing all symbols on the input tape.

Definition 3.15   Measure-once quantum finite-state automaton (MO-1QFA)   A quantum finite-state automaton (QFA) is a tuple $M = (\mathbb{C}^m, \Sigma, \Delta, |s_0\rangle, \mathcal{H}_F)$ where

- finite-dimensional Hilbert space $\mathbb{C}^m$ represents the set of states,

- Σ is an input alphabet,

- $\Delta : \Sigma \mapsto \mathrm{SU}(m)$ is a function assigning (special) unitary matrix to each element of Σ,

- $|s_{\mathrm{in}}\rangle \in \mathbb{C}^m$ is the initial state,

- $\mathcal{H}_{\mathrm{accept}} \subset \mathbb{C}^m$ is a subspace of accepted states.

The computation of the quantum finite-state machine starts in the initial state $|s_0\rangle$. In i-th computational step, the machine reads input symbol $\omega_i \in \Sigma$ and the current state is updated by applying the transformation $U_{\omega_i} = \Delta(\omega_i)$. We denote by U_ω a unitary matrix representing a transition for a sequence of letters $\omega = \omega_1 \omega_2 \ldots \omega_{|\omega|}$, i.e., $U_\omega = U_{\omega_{|\omega|}} \cdots U_{\omega_2} U_{\omega_1}$. Using this notation, the final state of the machine reads

$$s_{|\omega|} = U_\omega |s_0\rangle = U_{\omega_{|\omega|}} \cdots U_{\omega_2} U_{\omega_1} |s_0\rangle. \tag{3.4}$$

The probability that the input word is accepted is defined as

$$|P_{\text{accept}} U_\omega |s_0\rangle|^2, \tag{3.5}$$

where P_{accept} is a projection operator on the subspace $\mathcal{H}_{\text{accept}}$.

For a given automaton M, we can define the quantum language recognized by M as the function $f_M(\omega) : A^\star \mapsto [0, 1]$ defined as

$$f_M(\omega) = |P_{\text{accept}} U_\omega |s_0\rangle|^2. \tag{3.6}$$

In Section 3.1 we saw that the class of regular languages fully characterizes the computational power of classical (deterministic and nondeterministic) finite-state automata. Following this fact and using Definition 3.15 we can introduce the class of quantum regular languages [85].

Definition 3.16 Quantum regular language (RMO) Quantum regular language is a language recognized by a measure-once quantum finite automaton.

Quantum finite-state automaton has many properties analogous to the properties of classical finite-state automaton. In particular we have [85].

Lemma 3.17 Pumping lemma for quantum languages [85] *If f is a quantum regular language, then for any word ω and $\epsilon > 0$, there exists k such that*

$$\forall_{u,v} |f(u\omega^k v) - f(uv)| < \epsilon.$$

and if the automaton for f is n-dimensional, then

$$k < \frac{1}{(c\epsilon)^n}$$

for some constant c.

Roughly speaking, the above lemma states that, in the case of quantum regular languages, any word can by pumped.

3.2.2   MEASURE-MANY QUANTUM FINITE AUTOMATON

Measure-many quantum finite automaton (MM-QFA) provides an alternative approach to constructing quantum automata. First models of this type, namely the model of one-way measure-many quantum finite automaton (MM-1QFA) and two-way measure-many quantum finite automaton (MM-2QFA), were introduced by Kondacs and Wartous in [84] and subsequently studied in [86, 87].

The main difference between measure-many and measure-once quantum automata is the act, that the MM-QFA allows for the measurement to be an element of computational process, while MO-QFA allows the measurement to appear only as the final step of computation.

Definition 3.18   MM-1QFA   An MM-1QFA is a tuple $(Q, \Sigma, \Delta, |s_0\rangle; Q_{\text{acc}}, Q_{\text{rej}}, \#, \$)$ where

- Q is a finite set of states,

- σ is an input alphabet,

- $\Delta : \Sigma \mapsto SU(m)$ is a function assigning (special) unitary matrix to each element of Σ,

- $|s_0\rangle \in Q$ is a starting state,

- $Q_{\text{acc}} \subseteq Q$ and $Q_{\text{rej}} \subseteq Q$ are (disjoint) sets of accepting and rejecting,

- $\#$ and $\$$ are the start and the end markers, respectively.

The states belonging to $Q_{\text{acc}} \cup Q_{\text{rej}}$ are called halting states. By $Q_{\text{non}} = Q (Q_{\text{acc}} \cup Q_{\text{rej}})$ we denote the non-halting subspace. It is also usually assumed that $\# \notin \sigma$ and $\$ \notin \Sigma$.

The computation of the MM-1QFA is more complicated than in the case of MO-1QFA. As before, it starts in some initial superposition $|s_0\rangle$. Subsequently, the unitary operation Δ_{a_1} is applied, resulting in $|s_1\rangle = \Delta_a |s_0\rangle$. This state can be written as

$$|s_1\rangle = \sum_{q_i \in Q_{\text{acc}}} \alpha_i |q_i\rangle + \sum_{q_j \in Q_{\text{rej}}} \beta_j |q_j\rangle + \sum_{q_k \in Q_{\text{non}}} \gamma_k |q_k\rangle. \tag{3.7}$$

Now the state is measured and the automaton

- accepts the input with the probability $\sum_i |\alpha_i|^2$,

- rejects the input with the probability $\sum_i |\beta_i|^2$,

- continues the computation with the probability $\sum_i |\gamma_i|^2$.

One should note that the measurement alters the state of the machine. In the next step of the computation the operation Δ_{a_2} is applied to the resulting state.

The transition function in Definition 3.18 describes the transition between internal quantum states. It can be written as

$$\Delta : Q \times \Sigma \times Q \mapsto \mathbb{C} \tag{3.8}$$

and describes, for each initial state q_i input symbol s_i, describes the amplitude of moving to the resulting state q_{i+1}.

In Definition 3.18 we assume that after each transition the head scanning input symbols moves to the right and reads the next symbol. If we allow the scanning head to move left, right, or not to move at all the transition function takes the form

$$\Delta : Q \times \Sigma \times Q \times \{\leftarrow, \downarrow, \rightarrow\} \mapsto \mathbb{C} \tag{3.9}$$

and, as in Definition 3.3, symbols $\{\leftarrow, \downarrow, \rightarrow\}$ describe the movement of the scanning head.

Definition 3.19 MM-2QFA   An MM-2QFA is defined as in the MM-1QFA with the exception of the transition function which in this case is defined according to Eq. 3.9.

The difference between MM-1QFA and MM-2QFA is similar as in the case of classical automata. In the quantum case, however, this difference leads to a greater computational power.

3.3   QUANTUM LANGUAGES

Using quantum automata one can extend the definition of regular languages to the quantum domain. It also possible to study the relations between different classes of languages and, in particular, describe relations between classical and quantum languages.

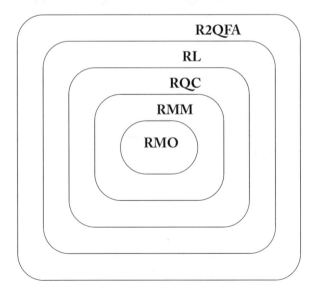

Figure 3.2: The relation between various classes of quantum languages. The outermost proper inclusion, **R2QFA** $\subsetneq$ **R2QFA**, reflects the fact that there exist non-regular languages which can be recognized by 2-way quantum finite automata.

The most important fact in this area is related to the increasing computational power of 2-way quantum automata. As we have seen, in the classical case the class of languages recognized by 2DFA is exactly the class of regular languages.

Kondacs and Watrous proved that 2-way quantum automata can recognize any regular language. Moreover, they gave an example of a language which is not regular and can be recognized by 2QFA [84].

Theorem 3.20   Kondacs and Watrous [84]   *Every regular language is accepted by some 2-way quantum finite automaton. Moreover, the non-regular language $\{a^n b^n : n \in \mathbb{N}\}$ can be recognized by some 2QFA with one-side error probability in the linear time.*

This shows that the quantum automata provide us with greater computational power compared to the classical finite automata.

The important classes of quantum languages include:

- **RMO** – languages recognized by 1-way measure-once quantum finite automata (MO-1QFA),

- **RMM** – languages recognized by 1-way measure-many quantum finite automata (MM-1QFA),

- **RQC** – languages recognized by 1-way quantum finite automata with control language (CL-1QFA) [88],

- **R2QFA** – languages recognized by two-way measure-many quantum finite automata (2QFA).

The relation between these classes of languages and regular languages is illustrated in Figure 3.2.

The fact that quantum automata can recognize non-regular languages stimulated active research in this area. In particular, classes of languages recognized by 1-way quantum automata were described in [89]. Various properties of quantum languages were considered by many authors [90, 91, 92, 93, 94].

3.4   SUMMARY

We have introduced and briefly characterized quantum finite automata. These models provide examples of classical finite state machines, extended with the ability to operate on data encoded in quantum states [95]. The properties of these models of computation are still a very active field of study, and some of the recent developments in this area can be found in e.g., [96, 97].

Among the other models of finite state machines studied in quantum information theory one can also point out quantum push-down automata (QPDA) [85, 98] and sequential quantum machines (QSM) [99].

The most important characteristic of quantum automata is that they are able to recognize non-regular languages. This fact suggests that quantum machines can provide us with greater computational power when compared to classical machines. One should note, however, that classical and quantum automata provide a somewhat limited model of computation. For this reason one cannot judge about the potential of quantum computers or their advantages over the classical computers using only the models discussed in this chapter.

CHAPTER 4

Computational Circuits

After presenting the basic facts about Turing machines and quantum finite automata we are ready to introduce more user-friendly models of computing devices, namely the Boolean circuit model. This model is widely used in classical computer science, and the quantum version of this model is the *de facto* standard for describing quantum algorithms and protocols.

We start by introducing Boolean circuits used to describe the classical computational process. Next, we introduce reversible circuits and discuss the motivation for considering the reversible computation.

As the evolution of the closed quantum system is unitary, and thus reversible, reversible circuits allow us to introduce the quantum circuits model in a natural fashion. We extend reversible circuits to the quantum realm by introducing the quantum circuits model, sometimes referred to as the quantum gate array model. We also introduce elementary quantum gates used for describing quantum information processing and quantum algorithms.

The quantum circuits model [6, 100] is one of the most popular among scientists working in the field of quantum information theory. It is mainly used for representation of unitary gates. It can be also used, to some extent, to represent arbitrary quantum computational processes, including measurement. However, the limitation of the quantum circuits model in this area motivated the development in the field of quantum programming languages.

As we will see, this model does not allow us to express some elements useful in the programming of quantum computers. For example, it is impossible to express in this model quantum conditional structures available in many quantum programming languages (see: Chapters 5, 8, and 9). For this reason the quantum circuit model is limited as a quantum programming technique.

From the quantum programming point of view the quantum circuits model is important as it is commonly used to represent the intermediate phase of translation of the high-level program into its representation in terms of physically realizable instructions. In this case it is used in the quantum programming environment described in Chapter 6.

4.1   BOOLEAN CIRCUITS

The study of Boolean circuits goes back to the early papers of Shannon [13]. The rapid development in this area was stimulated by Shannon and Riordan [101], who used Boolean algebra to design and analyze switching circuits, and by Lupanov [102], who was working on the efficient synthesis of switching circuits.

Boolean circuits can operate on fixed-length input strings consisting of binary numbers. They are used to compute functions of the form

$$f : \{0, 1\}^m \mapsto \{0, 1\}^n. \tag{4.1}$$

Basic gates (functions) which can be used to define such circuits are:

- $\wedge : \{0, 1\}^2 \mapsto \{0, 1\}, \wedge(x, y) = 1 \Leftrightarrow x = y = 1$ (*logical and*),

- $\vee : \{0, 1\}^2 \mapsto \{0, 1\}, \vee(x, y) = 0 \Leftrightarrow x = y = 0$ (*logical or*),

- $\sim : \{0, 1\} \mapsto \{0, 1\}, \sim(x) = 1 - x$ (*logical not*).

The set of gates is called universal if all functions $\{0, 1\}^n \mapsto \{0, 1\}$ can be constructed using the gates from this set. It is easy to show that the set of functions composed of the $\sim$, $\vee$, and $\wedge$ is universal. Thus it is possible to compute any functions $\{0, 1\}^n \mapsto \{0, 1\}^m$ using only these functions. The full characteristic of universal sets of functions was given by Post in 1949 [103].

Using the above set of functions a Boolean circuit is defined as follows.

Definition 4.1 Boolean circuit   A Boolean circuit is an acyclic direct graph with nodes labelled by input variables, output variables, or logical gates $\vee$, $\wedge$, or $\sim$.

An input variable node has no incoming arrows while an output variable node has no outgoing arrows. The example of a Boolean circuit computing the sum of bits x_1 and x_2 is given in Figure 4.1.

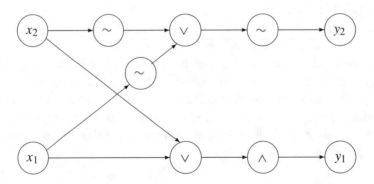

Figure 4.1: The example of a Boolean circuit computing the sum of bits x_1 and x_2 [100]. Nodes labelled x_1 and x_2 represent input variables and nodes labelled y_1 and y_2 represent output variables.

Note that in general it is possible to define a Boolean circuit using different sets of elementary functions. Since functions $\vee$, $\wedge$, and $\sim$ provide a universal set of gates we defined Boolean circuit using these particular functions.

Function $f : \{0, 1\}^m \mapsto \{0, 1\}$ is defined on the binary string of an arbitrary length. Let $f_n : \{0, 1\}^m \mapsto \{0, 1\}^n$ be a restriction of f to $\{0, 1\}^n$. For each such restriction there is a Boolean circuit C_n computing f_n. We say that $C_0, C_1, C_2, \ldots$ is a *family of Boolean circuits* computing f.

Any binary language $L \subset \{0, 1\}^*$ can be accepted by some family of circuits. However, since we need to know the value of f_n to construct a circuit C_n such family is not an algorithmic device at all. We can state that there exists a family accepting the language but we do not know how to build it [11].

To show how Boolean circuits are related to Turing machines we introduce uniformly generated circuits.[1]

Definition 4.2   We say that language $L \in A^*$ has uniformly polynomial circuits if there exists a Turing machine M that an input $\underbrace{1 \ldots 1}_{n}$ outputs the graph of circuit C_n using space $O(\log n)$, and the family $C_0, C_1, \ldots$ accepts L.

The following theorem provides a link between uniformly generated circuits and Turing machines.

Theorem 4.3   *A language L has uniformly polynomial circuit iff $L \in \mathbf{P}$.*

The quantum circuits model is analogous to uniformly polynomial circuits. They can be introduced as the straightforward generalization of reversible circuits.

4.2   REVERSIBLE CIRCUITS

The evolution of isolated quantum systems is described by a unitary operator U, i.e., an operator U such that $U U^\dagger = \mathbb{I}$ (see: e.g. [104]. The main difference with respect to classical evolution is that this type of evolution is always *reversible*. In other words, if we aim to describe a quantum computation by a Boolean circuit, such circuit has to be reversible.

For this reason before introducing quantum circuits we define a reversible Boolean circuit [105].

Definition 4.4   Reversible gate   A classical reversible function (gate) $\{0, 1\}^m \mapsto \{0, 1\}^m$ is a permutation.

One should note that in order for the classical function to be reversible, it has to have the same number of inputs and outputs. Moreover, as any reversible gate is a permutation, its output is completely determined by its input.

Definition 4.5   Reversible circuit   A reversible Boolean circuit is a Boolean circuit composed of reversible gates.

---

[1]We have already used this concept in Chapter 2.

The important fact, expressed by the following theorem, allows us to simulate any classical computation on a quantum machine described using a reversible circuit.

Theorem 4.6   Bennet [106]   *Any Boolean circuit can be simulated using a reversible Boolean circuit.*

The above fact is of great interest to computer scientists. This is motivated by the fact expressed in the Landauer's principle [107], which provides a smaller amount of energy required to erase the state of a bit.

Theorem 4.7   Landauer [107]   *If the machine erases one bit of information, the entropy of its environment is increased by at least $k_B \ln 2$, where k_B is Boltzmann constant.*

The above rule can be stated in terms of energy dissipated by the machine into the environment: by erasing the state of one bit, the machine dissipated at least $k_B T \ln 2$ amount of energy, where T is the temperature of the environment.

The energy is dissipated, however, only if the information is erased (i.e., destroyed). Therefore, if one processes the information using reversible gates only, it is in principle possible to perform computation without the loss of efficiency.

This principle can be understood as a simple consequence of the second law of thermodynamics [108]. One should note, however, that the Landauer's principle was among the most important factors motivating the consideration of computers operating on a basis of quantum mechanics.

4.2.1   UNIVERSAL REVERSIBLE GATES

Having defined reversible circuits, we can ask about the minimal set of reversible gates required to assemble any such circuit. As in the case of a non-reversible circuit one can introduce the universal set of functions for reversible circuits.

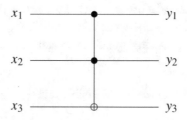

Figure 4.2: Classical Toffoli gate is universal for reversible circuits. It was also used in [109] to provide the universal set of quantum gates.

The important example of a gate universal for reversible Boolean circuits is a Toffoli gate. The graphical representation of this gate is presented in Figure 4.2. The following theorem was proved by Toffoli [110].

Theorem 4.8   *A Toffoli gate is a universal reversible gate.*

As we will see in the following section it is possible to introduce two-bit quantum gates which are universal for quantum circuits. This is impossible in the classical case and one needs at least a three-bit gate to construct the universal set of reversible gates.

In particular, any reversible circuit is automatically a quantum circuit. For a given reversible classical operation, its quantum counterpart is defined by the action on the base states. However, quantum circuits offer much more diversity in terms of the number of allowed operations.

4.3   QUANTUM CIRCUITS

The computational process of the quantum Turing machine is complicated since data as well as control variables can be in a superposition of base states. To provide a more convenient method of describing quantum algorithms one can use a quantum circuits model. This model is sometimes called a *quantum gate array* model.

The quantum circuits model was first introduced by Deutsch in [109] and it is the most commonly used notation for quantum algorithms. It is much easier to imagine than the quantum Turing machine since the control variables (executed steps and their number) are classical. There are only quantum data (e.g., qubits or qudits and unitary gates) in a quantum circuit.

A quantum circuit consists of the following elements (see Table 4.2):

- the finite sequence of *wires* representing qubits or sequences of qubits (quantum registers),

- quantum gates representing elementary operations from the particular set of operations implemented on a quantum machine,

- measurement gates representing a measurement operation, which is usually executed as the final step of a quantum algorithm. It is commonly assumed that it is possible to perform the measurement on each qubit in canonical basis $\{|0\rangle, |1\rangle\}$ which corresponds to the measurement of the S_z observable.

The concept of a quantum circuit is the natural generalization of acyclic logic circuits studied in classical computer science. Quantum gates have the same number of inputs as outputs. Each n-qubit quantum gate represents the 2^n-dimensional unitary operation of the group $SU(2^n)$, i.e., generalized rotation in a complex Hilbert space.

The main advantage of this model is its simplicity. It also provides very convenient representation of physical evolution in quantum systems.

From the mathematical point of view quantum gates are unitary matrices acting on n-dimensional Hilbert space. They represent the evolution of an isolated quantum system [6].

The problem of constructing new quantum algorithms requires more careful study of operations used in the quantum circuit model. In particular we are interested in efficient decomposition of quantum gates into elementary operations.

We start by providing the basic characteristics of unitary matrices. [6, 111]

Theorem 4.9   *Every unitary 2×2 matrix $G \in U(2)$ can be decomposed using elementary rotations as*

$$G = \Phi(\delta) R_z(\alpha) R_y(\theta) R_z(\beta) \tag{4.2}$$

where

$$\Phi(\xi) = \begin{pmatrix} e^{i\xi} & 0 \\ 0 & e^{i\xi} \end{pmatrix}, \quad R_y(\xi) = \begin{pmatrix} \cos(\xi/2) & \sin(\xi/2) \\ -\sin(\xi/2) & \cos(\xi/2) \end{pmatrix},$$

and

$$R_z(\xi) = \begin{pmatrix} e^{i\frac{\xi}{2}} & 0 \\ 0 & e^{-i\frac{\xi}{2}} \end{pmatrix}.$$

We introduce the definition of quantum gates as stated in [100].

Definition 4.10   A quantum gate U acting on m qubits is a unitary mapping on $\mathbb{C}^{2^m} \equiv \underbrace{\mathbb{C}^2 \otimes \ldots \otimes \mathbb{C}^2}_{m \text{ times}}$

$$U : \mathbb{C}^{2^m} \mapsto \mathbb{C}^{2^m}, \tag{4.3}$$

which operates on the fixed number of qubits.

Formally, a quantum circuit is defined as the unitary mapping which can be decomposed into the sequence of elementary gates.

Definition 4.11   A quantum circuit on m qubits is a unitary mapping on $\mathbb{C}^{2^m}$, which can be represented as a concatenation of a finite set of quantum gates.

Any reversible classical gate is also a quantum gate. In particular logical gate $\sim$ (negation) is represented by quantum gate NOT, which is realized by σ_x Pauli matrix.

As we know any Boolean circuit can be simulated by a reversible circuit and thus any function computed by a Boolean circuit can be computed using a quantum circuit. Since a quantum circuit operates on a vector in complex Hilbert space it allows for new operations typical for this model.

The first example of quantum gate which has no classical counterpart is $\sqrt{NOT}$ gate. It has the following property

$$\sqrt{NOT}\sqrt{NOT} = NOT, \tag{4.4}$$

which cannot be fulfilled by any classical Boolean function $\{0, 1\} \mapsto \{0, 1\}$. Gate $\sqrt{N}$ is represented by the unitary matrix

$$\sqrt{NOT} = \frac{1}{2} \begin{pmatrix} 1+i & 1-i \\ 1-i & 1+i \end{pmatrix}. \tag{4.5}$$

Table 4.1: Truth table for XOR gate, which is classical equivalent of the quantum CNOT gate.

x_1	x_2	x_1 XOR x_2
0	0	0
0	1	1
1	0	1
1	1	0

Another example is Hadamard gate H. This gate is used to introduce the superposition of base states. It acts on the base state as

$$H|0\rangle = \frac{1}{\sqrt{2}}\left(|0\rangle + |1\rangle\right), \; H|1\rangle = \frac{1}{\sqrt{2}}\left(|0\rangle - |1\rangle\right). \tag{4.6}$$

If the gate G is a quantum gate acting on one qubit it is possible to construct the family of operators acting on many qubits. The particularly important class of multiqubit operations is the class of controlled operations.

Definition 4.12 Controlled gate Let G be a 2×2 unitary matrix representing a quantum gate. Operator

$$|1\rangle\langle 1| \otimes G + |0\rangle\langle 0| \otimes \mathbb{I} \tag{4.7}$$

acting on two qubits, is called a controlled-G gate. Here $A \otimes B$ denotes the tensor product of gates (unitary operator) A and B, and $\mathbb{I}$ is an identity matrix. If in the above definition we take $G = NOT$ we get

$$|1\rangle\langle 1| \otimes \sigma_x + |0\rangle\langle 0| \otimes \mathbb{I} = \begin{pmatrix} 1 & 0 & 0 & 0 \\ 0 & 1 & 0 & 0 \\ 0 & 0 & 0 & 1 \\ 0 & 0 & 1 & 0 \end{pmatrix}, \tag{4.8}$$

which is the definition of $CNOT$ (controlled-NOT) gate. This gate can be used to construct the universal set of quantum gates. This gate also allows us to introduce entangled states during computation

$$CNOT(H \otimes \mathbb{I})|00\rangle = CNOT\frac{1}{\sqrt{2}}\left(|0\rangle + |1\rangle\right) \otimes |0\rangle = \frac{1}{\sqrt{2}}\left(|00\rangle + |11\rangle\right) \tag{4.9}$$

Quantum CNOT gate computes the value of x_1 XOR x_2 in the first register and stores values of x_2 in the second register. The classical counterpart of $CNOT$ gate is XOR gate.

Other examples of single-qubit and two-qubit quantum gates are presented in Table 4.2. In Figure 4.3 a quantum circuit for quantum Fourier transform on three qubits is presented.

Table 4.2: Basic gates used in quantum circuits with their graphical representation and mathematical form. Note that measurement gate is represented in Kraus form, since it is the example of non-unitary quantum evolution.

Name	Graphical representation	Mathematical form
Hadamard	$-\boxed{H}-$	$\frac{1}{\sqrt{2}}\begin{pmatrix}1 & 1\\ 1 & -1\end{pmatrix}$
Pauli X (σ_x, NOT)	$-\boxed{X}-$	$\begin{pmatrix}0 & 1\\ 1 & 0\end{pmatrix}$
Pauli Y (σ_y)	$-\boxed{Y}-$	$\begin{pmatrix}0 & -i\\ i & 0\end{pmatrix}$
Pauli Z (σ_z)	$-\boxed{Z}-$	$\begin{pmatrix}1 & 0\\ 0 & -1\end{pmatrix}$
Phase	$-\boxed{S}-$	$\begin{pmatrix}1 & 0\\ 0 & i\end{pmatrix}$
$\pi/8$	$-\boxed{T}-$	$\begin{pmatrix}1 & 0\\ 0 & e^{i\pi/4}\end{pmatrix}$
$R(\phi)$	$-\boxed{R_\phi}-$	$\begin{pmatrix}1 & 0\\ 0 & e^{i\phi}\end{pmatrix}$
CNOT		$\begin{pmatrix}1 & 0 & 0 & 0\\ 0 & 1 & 0 & 0\\ 0 & 0 & 0 & 1\\ 0 & 0 & 1 & 0\end{pmatrix}$
SWAP		$\begin{pmatrix}1 & 0 & 0 & 0\\ 0 & 0 & 1 & 0\\ 0 & 1 & 0 & 0\\ 0 & 0 & 0 & 1\end{pmatrix}$
Measurement	$-\boxed{\measuredangle}-$	$\left\{\begin{pmatrix}1 & 0\\ 0 & 0\end{pmatrix}, \begin{pmatrix}0 & 0\\ 0 & 1\end{pmatrix}\right\}$
qubit	——	wire $\equiv$ single qubit
n qubits	—/—	wire representing n qubits
classical bit		double wire $\equiv$ single bit

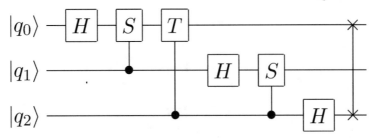

Figure 4.3: Quantum circuit representing quantum Fourier transform for three qubits. Elementary gates used in this circuit are described in Table 4.2.

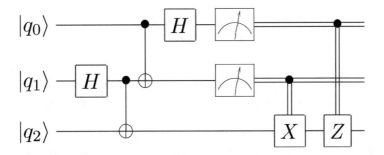

Figure 4.4: Circuit for quantum teleportation. Double lines represent the operation which is executed depending on the classical data obtained after the measurement on a subsystem.

One can extend Definition 4.12 and introduce quantum gates with many controlled qubits. This type of gate is very useful for expressing quantum algorithms operating on large quantum registers. As we will see in Chapter 5, the ability to express this type of gate in a straightforward manner is one of the typical features of quantum programming languages.

Definition 4.13 Let G be a 2×2 unitary matrix. Quantum gate defined as

$$| \underbrace{1 \ldots 1}_{n-1} \rangle \langle \underbrace{1 \ldots 1}_{n-1} | \otimes G + \sum_{l \neq \underbrace{1 \ldots 1}_{n-1}} |l\rangle \langle l| \otimes \mathbb{I} \qquad (4.10)$$

is called $(n-1)$-controlled G gate. We denote this gate by $\wedge_{n-1}(G)$.

This gate $\wedge_{n-1}(G)$ is sometimes referred to as a generalized Toffoli gate or a Toffoli gate with m controlled qubits. Graphical representation of this gate is presented in Figure 4.5.

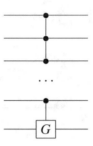

Figure 4.5: Generalised quantum Toffoli gate acting on n qubits. Gate G is controlled by the state of $n - 1$ qubits according to Definition 4.13.

The important feature of quantum circuits is expressed by the following universality property [111].

Theorem 4.14   *The set of gates consisting of all one-qubit gates $U(2)$ and one two-qubit CNOT gate is universal in the sense that any n-qubit operation can be expressed as the composition of these gates.*

Note that, in contrast to the classical case, where one needs at least three-bit gates to construct a universal set, quantum circuits can be simulated using one two-qubit universal gate.

In order to implement a quantum algorithm one has to decompose many qubit quantum gates into elementary gates. It has been shown that almost any n-qubit quantum gate ($n \geq 2$) can be used to build a universal set of gates [112] in the sense that any unitary operation on the arbitrary number of qubits can be expressed as the composition of gates from this set. In fact the set consisting of two-qubit exclusive-or (XOR) quantum gates and all single-qubit gates is also universal [111].

Let us assume that we have the set of gates containing only CNOT and one-qubit gates. In [113] a theoretical lower bound for the number of gates required to simulate a circuit using these gates was derived. The efficient method of elementary gates sequence synthesis for an arbitrary unitary gate was presented in [114].

Theorem 4.15   Shende-Markov-Bullock   *Almost all n-qubit operators cannot be simulated by a circuit with fewer than $\lceil \frac{1}{4}[4^n - 3n - 1] \rceil$ CNOT gates.*

In [115] the construction providing the efficient way of implementing arbitrary quantum gates was described. The resulting circuit has complexity $O(4^n)$ which coincides with lower bound from Theorem 4.15.

It is useful to provide more details about the special case, when one uses gates with many controlled qubits and one target qubit. The following results were proved in [111].

Theorem 4.16   *For any single-qubit gate U the gate $\wedge_{n-1}(U)$ can be simulated in terms of $\Theta(n^2)$ basic operations.*

In many situations it is useful to construct a circuit which approximates the required circuit. We say that quantum circuits approximate other circuits with accuracy ε if the distance (in terms of Euclidean norm) between unitary transformations associated with these circuits is at most ε [111].

Theorem 4.17 *For any single-qubit gate U and $\varepsilon > 0$ gate $\wedge_{n-1}(U)$ can be approximated with accuracy ε using $\Theta(n \log \frac{1}{\varepsilon})$ basic operations.*

Note that the efficient decomposition of a quantum circuit is crucial in the physical implementation of quantum information processing. In a specific implementation the decomposition can be optimized using the set of elementary gates specific for target architecture. CNOT gates are of big importance since they allow us to introduce entangled states during computation. It is also hard to physically realize CNOT gate since one needs to control physical interaction between qubits.

4.4 SUMMARY

This chapter has introduced the most important formalism used in the quantum information theory, namely the model of quantum circuits.

We have introduced reversible circuits motivated by the Landauer's principle. One should note, however, that the first reversible model of computation—reversible Turing Machine—was introduced independently by Lecerf [116] and Bennett [106].

We should also point out some limitations of the quantum circuits model. The first of them is related to the lack of means for using classical control structures within this model. In the case of quantum Turing machine, this issue was addressed by the introduction of the classically controlled quantum Turing machine (see: Section 2.4.2). The extension of this type is not feasible in the case of quantum circuits.

The second limitation stems from the need of feeding classical data to the quantum computer and reviving them as a result of the quantum computational process. In the language of quantum mechanics, the first process is described by the preparation of a state and the second as the measurement. Both operations naturally appear in the description of any quantum procedure and one of the strengths of quantum programming languages is the ability to easily incorporate them into the formal description of quantum programs.

CHAPTER 5

Random Access Machines

The main obstacle for writing quantum programs using the quantum circuits model is its lack of classical elements. Quantum circuits can be used to express quantum data and operations only and do not provide a mechanism which would allow to control the operations on quantum memory using a classical machine. However, in many quantum algorithms the classical and quantum elements are mixed together and usually only a part of computation is purely quantum. For this reason quantum algorithms are usually described using a mixture of mathematical representation, quantum circuits and classical algorithms.

The quantum random access machine is interesting for us since it provides a convenient model for developing quantum programming languages. However, these languages are our main area of interest. We see no point in providing the detailed description of this model as it is given in [117] together with the description of hybrid architecture used in quantum programming.

In this chapter we review a classical RAM model. We also introduce a Quantum Random Access Machine (QRAM) model. These models allows us to introduce a programming language used to write programs for QRAm, namely quantum pseudocode. We provide an introduction to the quantum pseudocode and describe how it can be used to express quantum information processing including classical elements.

5.1 CLASSICAL RAM MODEL

The classical model of Random Access Machine (RAM) is the example of a more general class of register machines [15, 16, 118]. The crucial feature provided by the RAM model is the ability of indirect addressing.

This model was introduced to provide the means for reasoning about the time and memory complexity using a model which resembles actual computing devices. Prior to the introduction of the model such results were stated using a multi-tape Turing machine model.

5.1.1 ELEMENTS OF THE MODEL

The RAM machine consists of an unbounded sequence of memory registers and a finite number of arithmetic registers. Each register may hold an arbitrary integer number.

Despite the difference in the construction between a Turing machine and RAM, it can be easily shown that a Turing machine can simulate any RAM machine with polynomial slow-down only [11].

Definition 5.1   Random Access Machine   A random access machine consists of

- an array or registers $R = (r_0, r, \ldots, r_n)$, which can be used to store integer numbers,

- a program counter or a state register K, which identifies the current instruction to be executed,

- a set of allowed instructions Π.

The program for the RAM is a finite sequence of instructions $\Pi = (\pi_1, \ldots, \pi_n)$. At each step of execution register i holds an integer r_i and the machine executes instruction π_κ, where κ is the value of the program counter. Arithmetic operations are allowed to compute the address of a memory register.

5.1.2   RAM-ALGOL

For the purpose of describing programs for the RAM machine a RAM-ALGOL programming language was introduced [15]. RAM-ALGOL was defined as a subset of ALGOL [119], with the following limitations

- only integer numbers are allowed and real numbers were eliminated,

- arithmetic operations are limited to $+$ and $-$,

- procedures are not allowed to be recursive,

- arrays are one-dimensional.

As instructions for classical RAM machine can be described in ALGOL-like language or classical pseudocode, instructions for the QRAM machines are described using the quantum pseudocode. However, the syntax of existing quantum programming languages was also influenced by the languages of the ALGOL family.

Besides these limitations the RAM-ALGOL is a full-featured programming language. This demonstrates that the RAM model allows us to introduce the concept of programming language in a very straightforward manner.

It is worth noting that programming languages can be defined without using the RAM model. An interesting programming language for a Turing machine $\mathcal{P}''$, providing the minimal set of instructions, was introduced by Böhm in [21].

5.2   QUANTUM RAM MODEL

The quantum random access machine (QRAM) model is the extension of the classical RAM. The basic idea behind this model is that it allows us to perform quantum computation controlled by a classical device. Thus the basic difference between the classical RAM and its quantum version is the ability to operate on quantum data.

QRAM can exploit quantum resources and, at the same time, can be used to perform any kind of classical computation. It allows us to control operations performed on quantum registers and provides the set of instructions for defining them.

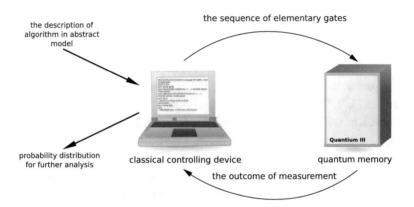

Figure 5.1: The model of classically controlled quantum machine [117]. Classical computer is responsible for performing unitary operations on quantum memory. The results of quantum computation are received in the form of measurement results.

Quantum circuits, introduced in Chapter 4, provide a powerful, formalized method for describing the quantum computational process occurring in an isolated system. In such situations the evolution of a quantum system is described by a unitary operator.

However, at the final step of computation, the physical system used to run a quantum program undergoes a non-unitary evolution—a measurement. Moreover, the results of this measurement can be used in the next phase of the program. Thus these results are used e.g., to prepare the initial state of quantum memory or as parameters of the unitary evolution.

Clearly quantum circuits cannot be used to represent classical control of the operations on quantum memory.

The model of Quantum Random Access Machine (QRAM) was developed in order to address this problem [120]. The QRAM model is built on the assumption that the quantum computer has to be controlled by a classical device [117]. Schematic presentation of such architecture is provided in Figure 5.1.

The quantum part of the QRAM model is used to generate probability distribution. This is achieved by performing measurement on quantum registers. The obtained probability distribution has to be analyzed using a classical computer.

5.3   QUANTUM PSEUDOCODE

Quantum algorithms are, in most of the cases, described using the mixture of quantum gates, mathematical formulas and classical algorithms. The first attempt to provide a uniform method of describing quantum algorithms was made in [121], where the author introduced a high-level notation based on the notation known from computer science textbooks [122].

In [120] the first formalized language for the description of quantum algorithms was introduced. Moreover, it was tightly connected with the model of quantum machine called quantum random access machine (QRAM).

Quantum pseudocode proposed by Knill [120] is based on conventions for classical pseudocode proposed in [122, Chapter 1]. Classical pseudocode was designed to be readable by professional programmers, as well as people who had done a little programming. Quantum pseudocode introduces operations on quantum registers. It also allows us to distinguish between classical and quantum registers.

The most important element of the quantum pseudocode is the introduction of quantum registers. Quantum registers are distinguished by underlining them. They can be introduced by applying quantum operations to classical registers or by calling a subroutine which returns a quantum state. In order to convert a quantum register into a classical register the measurement operation has to be performed.

5.3.1   ELEMENTS OF QUANTUM PSEUDOCODE

Quantum pseudocode provides several methods for allocating and processing quantum registers.

Using the quantum pseudocode one can easily introduce a classical control of the quantum operations by using if/then statement

```
if a then
      Y(b)
```

where Y denotes σ_y operation.

The example of a program written in quantum pseudocode is presented in Listing 5.1. It shows the main advantage of the QRAM model over the quantum circuits model—the ability to incorporate classical control into the description of quantum algorithm.

One should note that single qubits building quantum registers can be addressed by using lower indices. Using these methods one can perform a quantum operation on a single qubit.

Pseudocode	Description
$\underline{a} \leftarrow 0^{\frown 5}$	allocation of a quantum register, consisting of 5 qubits, in the state $|00000\rangle$
$\underline{a} \leftarrow a$	conversion of a classical register a to a quantum register $\underline{a}$
$\underline{a} \leftarrow 5$	initialization of a quantum register $\underline{a}$ with an integer number
$\underline{a} \leftarrow \text{InitQReg}(x)$	execute a classical operation affecting the state of a quantum register $\underline{a}$
$\underline{a} \leftarrow \text{QProc}(\underline{b}, c)$	execute an operation affecting a state of $\underline{a}$, based on the quantum state of $\underline{b}$ and some classical data stored in c
$b \leftarrow 5$	declaration of a classical register containing the integer 5
$b \leftarrow \text{CProc}()$	store the result of the operation in a classical register b

Table 5.1: Syntax and semantics of quantum pseudocode.

Procedure: Fourier$(\underline{a}, d)$
Input: A quantum register $\underline{a}$ with d qubits numbered from 0 to $d-1$.
Output: The amplitudes of $\underline{a}$ are Fourier-transformed over $\mathbb{Z}_{2^d}$.

```
C: assign value to classical variable
ω ← e^{i2π/2^d}
C: perform sequence of gates
for  i = d-1 to i = 0
   for j = d-1 to j = i+1
   if a_j then R_{ω^{2^{d-i-1+j}}}(a_i)
     C: number of loops executing phase
     C: depends on the required accuracy
     C: of the procedure
   H(a_i)

C: change the order of qubits
for j = 0 to j = d/2 - 1
   SWAP(a_j, a_{d-a-j})
```

Listing 5.1: Quantum pseudocode for quantum Fourier transform (QFT) on d qubits [120]. Quantum circuit for this operation with $d = 3$ is presented in Figure 4.3 in Chapter 4.

Thanks to the possibility of addressing single qubits, quantum pseudocode can be used with different sets of elementary gates. This is crucial when we aim to decompose a multiqubit gate into elementary gates used on a particular target architecture. The set of available gates can differ on various physical realizations of quantum computing due to the implementation-specific physical constrains.

Operation $\mathcal{H}(\underline{a_i})$ executes a quantum Hadamard gate on a quantum register $\underline{a_i}$ and $\mathcal{SWAP}(\underline{a_i}, \underline{a_j})$ performs SWAP gate between $\underline{a_i}$ and $\underline{a_j}$. Operation $\mathcal{R}_\phi(\underline{a_i})$ executes quantum gate $R(\phi)$ (see Table 4.2 in Chapter 4) on the quantum register $\underline{a_i}$.

5.3.2   QUANTUM CONDITIONS

Another interesting element introduced by the quantum pseudocode is the usage of quantum conditions. As in the case of quantum registers, these types of conditional constructions are distinguished by using underlines, **if** $\underline{a}$ **then** `Proc()`.

For example, using this construction, the controlled NOT gate can expressed as

```
if a then
        NOT(b)
```

and the controlled phase gate as

```
if a then
        Rϕ(b)
```

This construction is especially useful when we deal with quantum gates with a know decomposition into conditional gates (see Listing 5.2).

```
C: SWAP gate implemented using quantum conditions
SWAP(a,b) {
if a then
        NOT(b)
if b then
        NOT(a)
if a then
        NOT(b)
}
```

Listing 5.2: The realization of the $\mathcal{SWAP}$ gate, which can be realized as a composition of three CNOT gates, using quantum conditions.

Quantum conditions were implemented in QCL and QML quantum programming languages described in Chapters 8 and 9.

5.3.3   MEASUREMENT

Quantum data processed using the quantum memory (see Figure 5.1) have to be measured in the final step of computation for further processing in the classical controlling unit.

In the quantum pseudocode, the measurement of a quantum register can be indicated using an assignment $a_j \leftarrow \underline{a_j}$.

We have already mentioned the need for exchanging classical data between the quantum computer and some external classical machine is crucial for describing any useful computational process performed on a quantum computer. Quantum mechanics described the exchange of data between the quantum and classical realm in terms of state preparation and measurement. The ability to represent these operations in a formal manner is one of the main features introduced by the quantum pseudocode and, as we will see in the next chapters, is indispensable in any quantum programming language.

5.4   SUMMARY

In this chapter we have briefly introduced a Quantum Random Access Machine model used as a theoretical basis for most of the quantum programming languages. As we will see in Chapters 8 and 9, the instructions available in the quantum pseudocode are very similar to those available in many existing quantum programming languages.

Among models based on QRAM one can point out the model of sequential quantum random machine (SQRAM) introduced by Nagarajan, Papanikolaou, and Williams [123]. In particular, a set of classical and quantum instructions for SQRAM was proposed and the issues with the compilation of high-level languages for this model were discussed. Moreover the simulator of SQRAM machines was implemented. Similar work was reported in [124], where the QRAM model was used to synthesize a general-purpose quantum circuit in a hardware description language VHDL.

Physical implementations of quantum computing face many challenges, in particular in creating a robust and scalable architecture for implementing quantum information processing. Many of these issues, along with proposed solutions, are discussed in a comprehensive book by Metodi, Faruque, and Chong [9].

One should note that the RAM model, along with the register machine, the pointer machine, and the Random Access Stored Program (RASP) machine, is one of the common models used in classical computer science. In particular, the RASP model allows us to store its instructions in registers. As such the RASP model is an example of the von Neumann architecture. Among other interesting idealized models of computation one can point out abacus machines introduced by Lamberk [125] and Minsky [126] (see also [74]). The Abacus machine is, similarly to RAM, a variant of the counter machine.

CHAPTER 6

Quantum Programming Environment

Since the main aim of this book is to present the advantages and the limitations of high-level quantum programming languages, we need to explain how these languages are related to the quantum random access machine model (QRAM), which provides a realistic model of computation for describing quantum computing devices. More precisely we would like to know what the relation is between the high-level programming language and the physical realization of the quantum program and how the high-level description can be translated into a low-level one.

For this purpose we present the overview of an architecture for quantum programming tools proposed in [127]. This architecture is based on the QRAM model. It provides basic abstract concepts used in the quantum programming languages described in Chapter 7.

The architecture is composed of stages required to connect a high-level description of the quantum computation process with the low-level constructs. Physical realization of these constructs concerns the available set of gates and error correcting codes as well as the limitations of the target physical architecture.

From our point of view, the most interesting part of this architecture is the quantum assembly language (QASM). This language provides an example of a low-level quantum programming language, i.e., it operates on elementary data types (qubits) and allows us to use only a fixed set of operations (gates). At the same time, the QASM provides a universal language for describing quantum computation without resorting to a specific physical realisation.

As we are mainly interested in the programming concepts used in the described architecture, we also provide an example of how QASM can be used without using a specific high-level quantum programming language. This example presents how QASM can be used independently as a language for describing quantum circuits.

6.1 ARCHITECTURE COMPONENTS

Let us start by introducing the elements of the layered software architecture for quantum computing design tools introduced in [127]. This illustrates how a high-level description of a quantum algorithm is related to a low-level (physical) realization of the algorithm. The high-level description, which is provided as an input for the quantum compiler tools, is transformed into a technology-specific implementation of operations required to execute the input program.

The architecture proposed in [127, 128] is designed for transforming a high-level quantum programming language into the technology-specific implementation set of operations.

This architecture is composed of four layers or phases:

- **High-level programming language** providing high-level mechanisms for performing useful quantum computation; this language should be independent from particular physical implementation of quantum computing.

- **Compiler of this language** providing architecture-independent optimization; also compilation phase can be used to handle quantum error correction required to perform useful quantum computation.

- **Quantum assembly language (QASM)**—assembly language extended by the set of instructions used in the quantum circuit model.

- **Quantum physical operations language (QCPOL)**, which describes the execution of quantum program in a hardware-dependent way; it includes physical operations and it operates on the universal set of gates optimal for a given physical implementation.

Design concepts

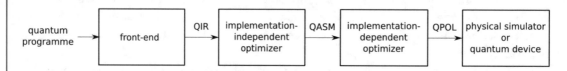

Abstract concepts

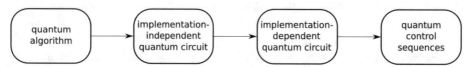

Figure 6.1: The most important concepts in the quantum programming architecture [127]. The upper diagram presents the elements of the compiler used to translate an abstract description of a quantum algorithm (quantum program) into an optimal sequence of controls used for controlling a physical device. The lower diagram presents the structures used in subsequent steps of the translation process. We start with the abstract representation in a high-level quantum programming language, which is transformed into a quantum circuit. Next, the implementation-independent circuit is optimized and transformed into a quantum circuit reflecting the limitations of the target physical device. In the last step, we obtain a sequence of instructions for classical devices responsible for controlling the target device.

The authors of [127, 128] do not define a specific high-level quantum programming language. They point out, however, that existing languages, mostly based on Dirac notation, do not provide the sufficient level of abstraction. They also stress, following [129], that it should have the basic set of features. We will discuss these basic requirements in detail in Chapter 7. At the moment quantum assembly language (QASM) is the most interesting part of this architecture, since it is tightly connected to the QRAM model.

It is also worth mentioning that a different approach for designing tools for linking high-level description with low-level concepts was presented by Bettelli *et al.* [129, 130]. In this approach the classical high-level language was extended with constructions incorporating elements required to express the quantum computational process. This architecture, however, does not provide a separation between different phases of translation and for this reason we will get back to this model in Chapter 8.

6.2   QUANTUM INTERMEDIATE REPRESENTATION

The first phase of translation of the high-level language into a physical description consists of the transformation of the input program into a quantum intermediate representation (QIR). In this form high-level concepts are translated into corresponding operations expressed in the language of abstract quantum mechanics. In particular,

- translation between quantum and classical data is achieved by quantum measurement,

- translation between classical and quantum data is expressed as state initialization,

- quantum conditions are expressed by gate multiplexing.

Moreover some high-level optimizations are applied.

6.3   QUANTUM ASSEMBLY LANGUAGE

Quantum assembly language used in the quantum programming environment should be powerful enough for representing high-level quantum programming language and it should allow for describing any quantum circuit. At the same time it must be implementation-independent so that it could be used to optimize the execution of the program with respect to different architectures.

As in the case of classical machines there are many possibilities of choosing a set of instructions in the definition of the quantum assembly languages. At the moment of writing, however, the most popular quantum assembly language is the one defined by Chuang [6, 131] for the purpose of graphical representation of quantum circuits using `qasm2circ` and used in [127] for the purpose of presenting the layered architecture introduced in the previous section. In the following we will refer to this language as QASM.

QASM uses qubits and cbits (classical bit) as basic units of information. As such it allows us to represent classically controlled quantum computing. Quantum operations consist of unitary operations and measurements. Moreover, each unitary operator is expressed in terms of single-qubit gates and CNOT gates.

```
# declare qubits
qubit    q0
qubit    q1
qubit    q2

# create EPR pair
h        q1
cnot     q1,q2
cnot     q0,q1

# Bell basis measurement
h        q0
nop      q1
measure q0
measure q1

# correction step
c-x      q1,q2
c-z      q0,q2
```

Listing 6.1: Description of the teleportation circuit in the QASM code. The QASM code as used in the quantum software architecture is used to represent the quantum program using only one-qubit quantum gates and CNOT gates. See Table 6.1 for the description of the operations used in this program.

In the architecture proposed in [127] each single-qubit operation is stored as the triple of rationals. Each rational multiplied by π represents one of three Euler-angles, which are sufficient to specify one-qubit operation.

One can note that only one three-qubit gate, namely the Toffoli gate, is introduced in Table 6.1. In the architecture introduced in [127], the quantum assembly language is required to be universal for representing any quantum program. To achieve this, only one-qubit and two-qubit gates are required. As such it does not have to provide an operation acting on three qubits.

QASM defined in [131], however, introduces the means for user-defined gates and gates operating on a greater number of qubits. These operators are presented in Table 6.2.

6.4   QUANTUM PHYSICAL OPERATIONS LANGUAGE

In this final phase of translation the QASM description of the quantum program is transformed into a Quantum Physical Operations Language (QPOL). At this stage the abstract description of the quantum program in terms of quantum gates is translated into an appropriate sequence of control pulses used to realize the quantum program on the physical device.

Table 6.1: Set of instructions defined in QASM [127, 131] along with accepted arguments and their descriptions. Most of the operations have an equivalent in one of the gates presented in Table 4.2 in Chapter 4. Arguments are described as labels associated with a qubit. Labels must be declared before their use, usually in the first section of the QASM program. In the case of operators accepting two or more arguments, the labels do not have to refer to the neighbouring qubits.

Instruction	Arguments	Description
qubit	ql,init	initialization of a qubit labelled with ql to the value init
cbit	cl,init	initialization of a classical bit labelled with cl to the value init
measure	ql	measurement on the qubit with the label ql and resulting in a classical bit, represented by a double wire
H, X, Y, Z, S or T	ql	one of the operations according to the description in Table 4.2.
c-x	cq,tq	controlled X gate
c-z	cq,tq	controlled Z gate
ZZ	b1,b2	synonym for c-z with a different graphical representation
SS	b1,b2	controlled S gate acting on qubits b1 and b2
swap	b1,b2	SWAP gate between qubits b1 and b2
space	ql	single qubit operator—an empty space on a register labelled with ql
cnot	cq,tq	two-qubit CNOT operation controlled by qubit cq with the target on tq
nop	ql	empty quantum operation (i.e., equivalent of a quantum wire)
zero	ql	re-initialization of a qubit ql to the state base $|0\rangle$
toffoli	cq1, cq2, tq	three-qubit Toffoli gate with two control qubits—cq1 and cq2—and the target qubit labelled by tq

Table 6.2: Elements of the QASM language allowing for the addition of multiqubit gates and user-defined operators. Element `Utwo` represent an arbitrary two-qubit gates, while elements `def` and `defbox` for user-defined gates can be represented graphically by a supplied symbol.

Instruction	Arguments	Description
`Utwo`	`ql1,ql2`	declaration of a two-qubit gate acting on qubits `ql1` and `ql2`
`slash`	`ql`	declaration of the multi-qubit compound register labelled by `ql`
`discard`	`ql`	operation indicating that the qubit `ql` will be ignored in the further part of the circuit which is equivalent to tracing-out the subsystem of the discarded qubit
`def`	`name`, `nctrl`, `sym`	a custom single-qubit operation `name` with an arbitrary number of controlling qubits given by `nctrl` and the graphical representation given by `sym`
`defbox`	`name`, `nbits`, `nctrl`, `texsym`	a custom operation `name` acting on `nbits` qubits, with an arbitrary number of controlling qubits given by `nctrl` and the graphical representation given by `sym`

The role of QPOL is to provide the means for expressing quantum procedures in the terms of instructions for a classical device controlling the physical quantum computer. Instructions available in QOPL can be roughly divided into six groups, namely

- initialization – preparation of the desired initial state,

- computation – preparation of control sequences required for the realization of the quantum gates used in the quantum program,

- measurement – read-out of the results by appropriate measurements,

- movement – control of the separation of the physical qubits required to switch on/off the collective evolution,

- classical computation – instructions required to process classical data in order to prepare different quantum gates using classical data,

- system specific instructions – instructions for controlling other physical degrees of freedom and aspects of functionality that do not fall into any of the above categories.

The need for incorporating into QOPL the ability to operate on classical data is motivated by the way the operations on quantum memory are related to the classical world. This is expressed in the QRAM model (see Figure 5.1 in Chapter 5), where the classical data obtained, as the result of the measurement, are used to prepare a set of instructions for operating on the quantum processor.

6.5 XML-BASED REPRESENTATION OF QUANTUM CIRCUITS

So far we have been dealing mainly with mathematical properties of quantum circuits. It is interesting, however, to see how quantum circuits can be represented and processed from the point of view of a programmer.

As we have seen in this chapter, the transformation of a high-level quantum program into a low-level language requires few steps and on each step a different method of describing the quantum program is used. In particular, quantum circuits are used in the middle phase of the translation. The crucial point in this phase is to provide a robust and scalable method for describing and processing quantum circuits.

The above motivated the development of a Quantum Markup Language (QML)[1], which provides a method of representing quantum circuits. QML can be used for simulating quantum computation, as well as for the purpose of executing a quantum program on a physical architecture.

QML was originally developed for the purpose of developing large-scale parallel simulations of quantum computing [134]. It was designed as a scalable and universal language for describing quantum circuits. It is based on eXtensible Markup Language (XML) and for this reason the QML description of quantum circuits can be easily processed by many modern computing systems.

Here we present the basic elements of Quantum Markup Language [134]. This shows that XML can be used for the representation of quantum gates in a very convenient way. QML can be used as an alternative method of description of quantum procedures replacing Quantum Assembly Language (QASM). As such it can be an element of the quantum programming environment introduced in Chapter 6.

6.5.1 BASIC ELEMENTS

A description of a quantum circuit in Quantum Markup Language (QML) document contains `<QML>` element, which is a parent for five elements, namely: `<Job>`, `<Circuit>`, `<GateLib>`, `<CircuitLib>` and `<Results>`.

An example of quantum circuit composed of basic quantum gates is presented in Listing 6.2. The description of a circuit consists of a sequence of `<Operation>` tags with an attribute `Step`. For each `<Operation>` tag a number of `<Application>` elements can be specified. Each `<Application>` specifies qubits to which it refers using `Bits` attribute and contains a reference to a quantum gate in the form of `<Gate>` tag.

---

[1]Not to be confused with the quantum functional programming language QML developed by Altenkirch and Grattage [132, 133], described in Chapter 9.

Graphical representation of the circuit defined in Listing 6.2 is presented in Figure 6.2. Such representation can be used with the help of software available from [135].

```xml
<?xml version="1.0"?>
<QML>
  <Circuit Name="test1" Id="simple.qml" Size="4" >
    <Operation Step="0">
    </Operation>
    <Operation Step="1">
      <Application Name="G" Bits="0,1">
        <Gate Type="CNOT"/>
      </Application>
      <Application Name="G" Bits="2">
        <Gate Type="HADAMARD"/>
      </Application>
      <Application Name="G" Bits="3">
        <Gate Type="HADAMARD"/>
      </Application>
    </Operation>
    <Operation Step="2">
      <Application Name="G" Bits="2,3">
        <Gate Type="CNOT"/>
      </Application>
    </Operation>
    <Operation Step='3'>
      <Application Bits='1'>
        <Gate Type="PHASE" Divisions="2"/>
      </Application>
    </Operation>
      <Operation Step='4'>
      <Application Bits='1,2,3'>
        <Gate Type="TOFFOLI"/>
      </Application>
    </Operation>
    </Circuit>
</QML>
```

Listing 6.2: Example of a QML document. This document contains only the <Circuit> section which defines the sequence of gates executed by the simulation engine. Also every gate presented here can be directly implemented by a unitary operation. This is not always true because QML allows for the conditional execution of gates.

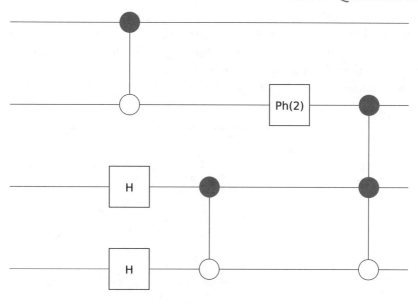

Figure 6.2: Simple circuit with controlled gates. Controlled qubits are marked with filled circles. SVG documents can use CSS for the specification of many attributes of graphical elements like color or line thickness. More examples can be found in [3].

Only the `<Circuit>` element is required for the definition of a quantum circuit. It contains the description of quantum circuits and it contains `<Operation>` elements, which represent the steps of the quantum algorithm. Attribute `Size` of the `<Circuit>` tag defines the number of qubits required for the execution of the presented circuit.

One can note that the bits on which gates should be performed are specified in the `<Operation>` tag. It is natural since the information required for the proper execution of gates is independent from their location. This can be used to include the parts of external descriptions of the circuit into the QML document and is similar to the mechanism of functions in procedural programming languages.

Since every popular programming language includes the support for XML it is easy to add the support for QML to any existing software for simulation of quantum computers and to integrate it with the existing simulation platform.

6.5.2 EXTERNAL CIRCUITS

One of the most interesting features of QML is the possibility of including external descriptions of circuits. This allows us to prepare parts of simulation using different tools and connect them using common XML format.

This feature also allows us to use any simulation engine supporting QML with programming languages such as QCL [136, 137] or any compiled high-level quantum programming language. Any such tool has to internally represent the input program in the form of elementary quantum gates. In order to use the external simulation engine one has to provide the conversion of the internal representation of quantum circuit into a QML description (see e.g., citeqml-zksi).

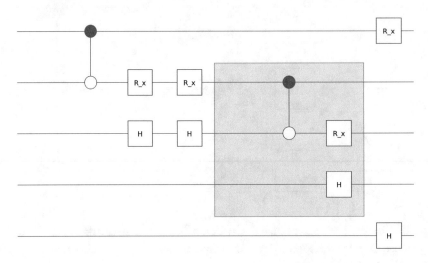

Figure 6.3: Circuit containing calls to other circuits. The group of gates in a frame is included in the circuit using <Circuit> element, which is a child of <Operation> element. It can also contain <Circuit> elements.

```
<?xml version="1.0"?>
<QML>
<Circuit Name="external" Id="external.qml" Size="5">
  <Operation Step="0">
  </Operation>
  <!-- ... -->
  <Operation Step='4'>
    <Application Bits='1,2,3'>
      <Gate Type="CIRCUIT"
      href="http://path/to/file.qml"/>
    </Application>
  </Operation>
  <!-- ... -->
</Circuit>
</QML>
```

Listing 6.3: QML document can contain `<Circuit>` tags with links to external definitions of gates. Such construction can be used to build a library of circuits and connect them dynamically during the execution. Attribute `href` of the tag `<Circuit>` contains URL of the file with the definition of the circuit.

Graphical representation of the circuit (see: Listing 6.3) containing reference to an external circuit is presented in Figure 6.3.

External elements of the circuit are included using `<Circuit>` element using its `href` attribute as it is presented in Listing 6.3. Since most of the programming languages support popular Internet protocols (*e.g.,* HTTP or FTP), it is easy to distribute the processing of the resulting QML document among many remote hosts.

6.6   SUMMARY

In this chapter we have presented a layered architecture designed for the purpose of translating a high-level programming language into low-level instructions suitable for an execution on a physical architecture used for implementing a quantum processing device. The main goal of the presented architecture is to divide a translation process into stages and thus allow us to use quantum programming languages independently on the given physical architecture. One should note, however, that the last stage of the process must take into account the features and limitations of the target architecture [9].

QASM language presented in this chapter can be, thanks to its ability to introduce user-defined sub-procedures, used successfully to describe general quantum algorithms and communication protocols. In many cases, however, such functionality is not needed. For this reason other variants of quantum assembly language were introduced [138] for the purpose of specialized quantum computing architectures.

The final phase of translation in the presented architecture is tightly connected with the physical architecture (implementation) of the quantum processing. As the construction of the efficient and robust physical architectures for quantum computing is still an active area of research [139, 140], one can expect a plethora of new fascinating problems arising during the concrete realizations of the abstract quantum compiler architecture.

CHAPTER 7

Quantum Programming Languages

In this chapter we introduce the last model of computation used in the theory of quantum information—quantum programming languages. The goal of this book is to provide a comprehensive introduction to high-level structures used in quantum information theory. For this reason we devote this and the next two chapters to the presentation of quantum programming languages.

We start by discussing the motivation for introducing and studying quantum programming languages. We argue that the need for the abstract, high-level description of quantum computation is needed in order to achieve a better understanding of the quantum computational process.

Next, we review basic requirements which should be satisfied by any useful quantum programming language. The list of requirements can be understood as complementary to the list of the requirements for any physical system to be useful as a quantum computer formulated by DiVincenzo and Loss [141].

We also compare features of the exiting quantum programming languages. A more detailed presentation of these languages is provided in Chapters 8 and 9. In this comparison we focus on languages which have a working interpreter and can be used as tools for experimenting with new quantum algorithms and protocols.

7.1 WHY STUDY QUANTUM PROGRAMMING LANGUAGES?

The models of computation introduced in the previous chapters allow us to express any quantum computational process. Each of these models is used in a specific area of quantum information. For example, quantum Turing machines and quantum automata are used mainly to study complexity of quantum algorithms. Quantum circuits, on the other hand, are used as the standard model of describing quantum algorithms and protocols.

The common feature of these models is that all of them are very closely related to the physical description of quantum computation in terms of qubits and unitary gates. To overcome this limitation, a significant amount of research has been devoted in order to create a model of quantum computation allowing us to express quantum computation using more abstract entities.

The first motivation for introducing and studying quantum programming languages stems from the limitations of the computational models we discussed in the previous chapters.

We have already mentioned during the presentation of the models of computation used in quantum information theory that some of them lack basic elements expected to be available in any fully functional quantum computing device. For example, the quantum circuits model lacks the ability to introduce classically controlled quantum computation.

This problem is addressed by the introduction of the QRAM model and quantum pseudocode, which can describe an interaction between quantum memory and classical computing device operating on this memory. This level of abstraction is implemented in the imperative programming languages described in Chapter 8. As we will see these languages provide a powerful tool for modelling quantum computation and testing new quantum algorithms.

The second motivation for the research effort in the area of quantum programming stemmed from the lack of abstraction available in the QRAM model and imperative programming languages. A program written in imperative languages provides a detailed description of computational steps required to execute the described algorithm. As we will see in the next chapter, the set of instructions available in most imperative quantum programming languages is more or less equivalent to the standard set of quantum gates used in most quantum algorithms (see Table 4.2 in Chapter 4).

This lack of abstraction motivated the development in the field of functional quantum programming languages. As we have already mentioned in Chapter 1, programming languages based on the functional paradigm aim to provide a description of *what* should be calculated by the program, instead of describing *how* this can be achieved.

7.2   QUANTUM PROGRAMMING BASICS

Quantum algorithms [32, 34, 142, 143] and communication protocols [25, 144, 145] are described using a language of quantum circuits [6]. While this method is convenient in the case of simple algorithms, it is very hard to operate on compound or abstract data types like arrays or integers using this notation [47, 146].

This lack of data types and control structures motivated the development of quantum pseudocode [120, 147] and various quantum programming languages [117, 127, 130, 148, 149, 150].

Several languages and formal models were proposed for the description of quantum computation process. The most popular of them is the quantum circuit model [109], which is tightly connected to the physical operations implemented in the laboratory. On the other hand the model of the quantum Turing machine is used for analyzing the complexity of quantum algorithms [49].

Another model used to describe quantum computers is the Quantum Random Access Machine (QRAM). In this model we have strictly distinguished the quantum part performing computation and the classical part, which is used to control computation. This model is used as the basis for most quantum programming languages [151, 152, 153]. Among high-level programming languages designed for quantum computers we can distinguish imperative and functional languages.

7.3   REQUIREMENTS FOR A QUANTUM PROGRAMMING LANGUAGE

We start by completing a wish list for a quantum programming language. Taking into account the drawbacks of the quantum circuit model described in Chapter 4 and elements available in the QRAM model described in Chapter 5, we can formulate basic requirements which have to be fulfilled by any quantum programming language [130].

- **Completeness:** Language must allow us to express any quantum circuit and thus enable the programmer to code every valid quantum program written as a quantum circuit.

- **Extensibility:** Language must include, as its subset, the language implementing some high level classical computing paradigm. This is important since some parts of quantum algorithms (for example Shor's algorithm) require nontrivial classical computation.

- **Separability:** Quantum and classical parts of the language should be separated. This allows us to execute any classical computation on a purely classical machine without using any quantum resources.

- **Expressivity:** Language has to provide high-level elements for facilitating the quantum algorithm's coding.

- **Independence:** The language must be independent from any particular physical implementation of a quantum machine. It should be possible to compile a given program for different architectures without introducing any changes in its source code.

As we will see, the languages presented in this Section fulfill most of the above requirements. The main problem is the *expressivity* requirement.

7.4   BASIC FEATURES OF EXISTING LANGUAGES

The earliest attempts for developing quantum programming languages are related to the great interest in quantum information science resulting from the spectacular algorithms by Shor. This research resulted in the development of programming languages oriented to the simulation of quantum algorithms. Among the languages from this category presented below we point out to QCL and QPL.

During the last decade, however, it became clear that the ability of constructing a fully-functional quantum computer is a very challenging task. Additionally, the most interesting results in quantum information theory are related to the use of quantum states for the purpose of communicating between distant parties. This is evidenced by the development of quantum programming languages focused on the modelling of quantum communication protocols. These languages are represented below by LanQ and cQPL.

7.4.1   IMPERATIVE LANGUAGES

Simulation-Oriented Approach

At the moment of writing this book the most advanced imperative quantum programming language is Quantum Computation Language (QCL) designed and implemented by Ömer [117, 136, 137]. QCL is based on the syntax of the C programming language and provides many elements known from classical programming languages. The interpreter is implemented using a simulation library for executing quantum programs on classical computers, but it can be in principle used as a code generator for a classical machine controlling a quantum circuit.

Along with QCL several other imperative quantum programming languages were proposed. Notably Q Language developed by Betteli [129, 130] and libquantum [154] have the ability to simulate noisy environment. Thus, they can be used to study decoherence and analyze the impact of imperfections in quantum systems on the accuracy of quantum algorithms.

Q Language is implemented as a class library for C++ programming language and libquantum is implemented as a C programming language library. Q Language provides classes for basic quantum operations like QHadamard, QFourier, QNot, QSwap, which are derived from the base class Qop. New operators can be defined using C++ class mechanism. Both Q Language and libquantum share some limitations with QCL, since it is possible to operate on single qubits or quantum registers (i.e., arrays of qubits) only.

In a similar fashion the basic high-level structures used for developing quantum programming languages were developed as a set of functions for the general purpose scientific computing system. The structures introduced in [4] are similar to the elements used in QCL and Q Language and were described in quantum pseudo-code based on QCL quantum programming language. They were implemented in GNU Octave language for scientific computing. The procedures used in the implementation are available as a package `quantum-octave` providing a library of functions, which facilitates the simulation of quantum computing. This package also allows us to incorporate high-level programming concepts into the simulation in GNU Octave.[1] As such it connects the features unique for high-level quantum programming languages, with the full palette of efficient computational routines commonly available in modern scientific computing systems.

Communication-oriented approach

Concerning the problems with physical implementations of quantum computers, it became clear that one needs to take quantum errors into account when modelling the quantum computational process. Also quantum communication has become a very promising application of quantum information theory over the last few years. Both facts are reflected in the design of new quantum programming languages.

LanQ developed by Mlnařík was defined in [150, 155]. It provides syntax based on C programming language. LanQ provides several mechanisms such as the creation of a new process by forking and interprocess communication, which support the implementation of multi-party proto-

---

[1] As the GNU Octave environment is to a large extent compatible with Matlab, the described package can also be used with Matlab.

cols. Moreover, operational semantics of LanQ has been defined. Thus, it can be used for the formal reasoning about quantum algorithms.

7.4.2   FUNCTIONAL LANGUAGES

The second group of quantum programming languages consists of the languages which are based on the functional paradigm.

Simulation-Oriented Approach

The use of a functional-based approach in the area of quantum information was motivated by the need for describing and modelling quantum objects using an abstract, formal language. Among the first attempts to bring a functional style of programming into the quantum domain we can point out the extension of lambda calculus introduced by Maymin [156]. The next attempt was by Tonder [157], who introduced quantum lambda calculus. It was introduced in a form of simulation library for Scheme programming language. Another interesting attempt was by Karczmarczuk, who defined and developed an abstract framework based on functional programming language for presenting the structures used in quantum mechanics [158]. This abstract framework was formulated using the Haskell programming language.

QPL [159] was the first functional quantum programming language. This language is statically typed and allows us to detect errors at compile-time rather than run-time. However, there is no working interpreter of QPL.

Communication-Oriented Approach

A more mature version of QPL is cQPL—communication capable QPL [148]. cQPL was created to facilitate the development of new quantum communication protocols. Its interpreter uses QCL as a back-end language so cQPL programs are translated into C++ code using QCL simulation library.

Table 7.1 contains the comparison of several quantum programming languages. It includes the most important features of existing languages. In particular we list the underlying mathematical model (i.e., pure or mixed states) and the support for quantum communication.

All languages listed in Table 7.1 are universal and thus they can be used to compute any function computable on a quantum Turing machine. Consequently, all these languages provide the model of quantum computation which is equivalent to the model of a quantum Turing machine.

In the following two chapters we compare the selected quantum programming languages and provide some examples of quantum algorithms and protocols implemented in these languages. We also describe their main advantages and limitations. We introduce the basic syntax of four of the languages listed in Table 7.1—QCL, LanQ, cQPL, and QML. This is motivated by the fact that these languages have a working interpreter and can be used to perform the simulations of quantum algorithms. We introduce basic elements of these languages required to understand basic programs. We also compare the main features of the presented languages.

Table 7.1: The comparison of quantum programming languages with information about implementation and basic features. This table is partially based on information from [148] and [155].

	QCL	Q Language	quantum-octave	LanQ	QPL	cQPL	QML
reference	[136]	[130]	[4]	[159]	[148]	[155]	[160]
implemented	✓	✓	✓	✓	–	✓	✓
semantics	–	–	✓	✓	✓	✓	✓
communication	–	–	–	✓	–	✓	–
universal	✓	✓	✓	✓	✓	✓	✓
mixed states	–	–	✓	✓	✓	✓	–

The main problem with current quantum programming languages is that they tend to operate on very low-level structures only. In QCL quantum memory can be accessed using only the `qreg` data type, which represents the array of qubits. In the syntax of cQPL data type `qint` has been introduced, but it is only synonymous for the array of 16 qubits. A similar situation exists in LanQ [155], where quantum data types are introduced using `q`n`it` keyword, where n represents a dimension of elementary unit (e.g., for qubits $n = 2$, for qutrits $n = 3$). However, only unitary evolution and measurement can be performed on variables defined using one of these types.

7.5 SUMMARY

During the last few years many quantum programming languages have been proposed and there are some papers presenting an overview of the development in this field [151, 152, 153, 161].

Quantum programming languages are commonly used as tools in the simulation of quantum computing. There are, however, many tools developed specifically for this purpose, which are usually based on commonly used scientific computing systems. The most up-to-date list of quantum computing simulators can be found at [162].

Most of the languages are developed as a proof of principle tools only. Also in most cases there are no working interpreters available. Even if the language was implemented, in many cases the interpreters are in a very preliminary stage.

It it also worth mentioning a considerable research effort for using an abstract approach based on the category theory for reasoning about the properties of quantum information processing protocols and for constructing a new quantum programming languages. Among the most important developments in this area one can point to works by Abramsky and Coecke [163, 164, 165] and by Selinger [166]. These studies have had a considerable impact on the development of functional quantum programming languages, which are discussed in Chapter 9.

CHAPTER 8

Imperative quantum programming

In this chapter we focus on quantum programming languages which are based on the imperative paradigm. The main characteristic of these languages is that they provide an exact description of the computational steps required to execute a quantum procedure.

Quantum programming languages in this family include Quantum Computation Language (QCL) created by Ömer [117, 136, 137] and LanQ developed by Mlnařík [149, 150, 155]. As our goal is to acquaint the reader with the quantum programming, we describe the languages for which there exists a working interpreter (see: [167] and [168]).

Imperative programming languages follow the design of the quantum computer proposed by the Quantum Random Access Machine model and are tightly connected with the quantum pseudocode presented in Chapter 5. This is especially true for QCL. As we will see QCL provides all elements introduced in quantum pseudocode. Moreover, it provides some elements important from the execution speed point of view.

Development of LanQ, on the other hand, was motivated by the lack of elements supporting the simulation of quantum protocols in the existing languages. For this reason its definition provides the user with some elements not introduced in quantum pseudocode, but crucial for developing programs for simulating quantum communication protocols like quantum teleportation or quantum direct communication.

We start by introducing basic elements of QCL, which is one of the most popular quantum programming languages. This language owes its popularity to the syntax resembling the syntax of many classical programming languages and for the offered speed of execution of quantum programs.

Next, we introduce the basic elements of LanQ. This language provides the support for quantum protocols. As we have already pointed out in Chapter 7, this fact reflects the recent progress in quantum information theory related to the application of quantum mechanics for transmitting information. The lack of support of quantum communication primitives in QCL does not mean that it is impossible to simulate quantum protocols in this language. The program for describing such simulations would be more cumbersome in QCL than in LanQ.

8.1 QCL

We begin our survey of imperative languages with QCL (Quantum Computation Language) [117, 136, 137]. At the moment of writing, QCL is the most advanced implemented quantum programming language. Its syntax resembles the syntax of C programming language [56] and classical data types are similar to data types in C or Pascal.

The programs written in QCL can be executed using the available interpreter [167]. The interpreter can be executed in a batch mode or in an interactive program. The interpreter is built on a top of `libqc` simulation library written in C++ and offers an excellent speed of execution of simulated programs. As the simulation of quantum computing requires a considerable amount of computing resources, there were also some attempts to provide a paralellized version of `libqc` library [169].

8.1.1   BASIC ELEMENTS

The basic built-in quantum data type in QCL is `qureg` (quantum register). It can be interpreted as the array of qubits (quantum bits).

```
qureg x1[2]; // 2-qubit quantum register x1
qureg x2[2]; // 2-qubit quantum register x2
H(x1); // Hadamard operation on x1
H(x2[1]); // Hadamard operation on the second qubit of the x2
```

Listing 8.1: Basic operations on quantum registers and subregisters in QCL.

QCL standard library provides standard quantum operators used in quantum algorithms, such as:

- Hadamard H and Not operations on many qubits,

- controlled not CNot with many target qubits and Swap gate,

- rotations: RotX, RotY and RotZ,

- phase Phase and controlled phase CPhase.

Most of them are described in Table 4.2 in Section 5.2.

Since QCL interpreter uses `qlib` simulation library, it is possible to observe the internal state of the quantum machine during the execution of quantum programs. The following sequence of commands defines two-qubit registers a and b and executes H and CNot gates on these registers.

```
qcl> qureg a[2];
qcl> qureg b[2];
qcl> H(a);
[4/32] 0.5 |0,0> + 0.5 |1,0> + 0.5 |2,0> + 0.5 |3,0>
qcl> dump
: STATE: 4 / 32 qubits allocated, 28 / 32 qubits free
0.5 |0> + 0.5 |1> + 0.5 |2> + 0.5 |3>
qcl> CNot(a[1],b)
```

```
[4/32] 0.5 |0,0> + 0.5 |1,0> + 0.5 |2,0> + 0.5 |3,0>
qcl> dump
: STATE: 4 / 32 qubits allocated, 28 / 32 qubits free
0.5 |0> + 0.5 |1> + 0.5 |2> + 0.5 |3>
```

Using dump command it is possible to inspect the internal state of the quantum computer. This can be helpful for checking if our algorithm changes the state of the quantum computer in the requested way.

One should note that dump operation is different from measurement, since it does not influence the state of the quantum machine. This operation can be realized using the simulator only.

8.1.2 QUANTUM MEMORY MANAGEMENT

Quantum memory can be controlled using quantum types qureg, quconst, quvoid, and quscratch. Type qureg is used as a base type for general quantum registers. Other types allow for the optimization of a generated quantum circuit. The summary of the types defined in QCL is presented in Table 8.1.

Table 8.1: Types of quantum registers used for memory management in QCL.

Type	Description	Usage
qureg	general quantum register	basic type
quvoid	register which has to be empty when operator is called	target register
quconst	register which must be invariant for all operators used in quantum conditions	quantum conditions
quscratch	register which has to be empty before and after the operator is called	temporary registers

8.1.3 CLASSICAL AND QUANTUM PROCEDURES AND FUNCTIONS

QCL supports user-defined operators and functions known from languages like C or Pascal. Classical subroutines are defined using procedure keyword. Also standard elements, known from C programming language, like looping (e.g., for i=1 to n { ... }) and conditional structures (e.g., if x==0 { ... }), can be used to control the execution of quantum and classical elements. In addition to this, it provides two types of quantum subroutines.

The first type is used for unitary operators. By using it one can define new operations, which in turn can be used to manipulate quantum data. For example operator diffuse, defined in Listing 8.2, defines *inverse about the mean* operator used in Grover's algorithm [34]. This allows us to define algorithms on the higher level of abstraction and extend the library of functions available for a programmer.

```
operator diffuse(qureg q) {
  H(q);                  // Hadamard Transform
  Not(q);                // Invert q
  CPhase(pi,q);          // Rotate if q=1111..
  !Not(q);               // undo inversion
  !H(q);                 // undo Hadamard Transform
}
```

Listing 8.2: The implementation of the inverse about the mean operation in QCL [117]. Constant `pi` represents number π. Exclamation mark ! is used to indicate that the interpreter should use the inverse of a given operator. Operation `diffuse` is used in the quantum search algorithm [34].

Using subroutines it is easy to describe quantum algorithms. Figure 8.1 presents QCL implementation of Deutsch's algorithm, along with the quantum circuit for this algorithm. This simple algorithm uses all the main elements of QCL. It also illustrates all the main ingredients of existing quantum algorithms.

The second type of quantum subroutine is called a *quantum function*.[1] It can be defined using the `qufunct` keyword. The subroutine of type `qufunct` is used for all transformations of the form

$$|n\rangle = |f(n)\rangle, \qquad (8.1)$$

where $|n\rangle$ is a base state and f is a one-to-one Boolean function. The example of quantum function is presented in Listing 8.4.

8.1.4   QUANTUM CONDITIONS

QCL introduces *quantum conditional statements*, i.e., conditional constructions where the quantum state can be used as a condition. Construction of this type was already introduced in the quantum pseudocode in Chapter 5. QCL, however, uses the same syntax for classical as well as quantum conditions.

QCL, as well as many classical programming languages, provides the conditional construction of the form

```
if be then
   block
```

where *be* is a Boolean expression and *block* is a sequence of statements.

QCL provides the means for using quantum variables as conditions. Instead of a classical Boolean variable, the variable used in condition can be a quantum register.

```
qureg a[2];
qureg b[2];
```

---

[1]Quantum functions are also called *pseudo-classic operators*.

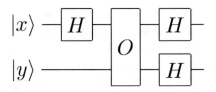

```
operator U(qureg x,qureg y) {
  H(x);
  Oracle(x,y);
  H(x & y);
}

procedure deutsch() {          // Classical control structure
  qureg x[1];                  // allocate 2 qubits
  qureg y[1];
  int m;
  {                            // evaluation loop
    reset;                     //   initialize machine state
    U(x,y);                    //   do unitary computation
    measure y,m;               //   measure 2nd register
  } until m==1;                // value in 1st register valid?
  measure x,m;                 // measure 1st register which
  print "g(0) xor g(1) =",m;   //   contains g(0) xor g(1)
  reset;                       // clean up
}
```

Figure 8.1: Quantum circuit for Deutsch's algorithm and QCL implementation of this algorithm (see [117] for more examples). Evaluation loop is composed of preparation (performed by reset instruction), unitary evolution (U(x,y) operator) and measurement. Subroutine Oracle() implements the function used in Deutsch's algorithm [20, 170].

```
// the sequence of statements
// ...
// perform CNot if a=|1...1>
if a {
  CNot(b[0], b[1]);
}
```

Listing 8.3: Example of a quantum conditional statement in QCL.

In this situation QCL interpreter builds and executes the sequence of $CNOT$ gates equivalent to the above condition. Here register a is called *enable register*.

In addition, quantum conditional structures can be used in quantum subroutines. Quantum operators and functions can be declared as conditional using cond keyword. For example

```
// conditional phase gate
extern cond operator Phase(real phi);
// conditional not gate
extern cond qufunct Not(qureg q);
```

declares a conditional Phase gate and a controlled NOT gate. Keyword extern indicates that the definition of a subroutine is specified in an external file. The enable register (i.e., quantum condition) is passed as an implicit parameter if the operator is used within the body of a quantum if-statement.

```
cond qufunct inc(qureg x) { // increment register
  int i;
  for i = #x-1 to 0 step -1 {
    CNot(x[i],x[0::i]);       // apply controlled-not from
  }                           // MSB to LSB
}


// equivalent implementation with constant enable register
qufunct cinc(qureg x,quconst e) { // Conditional increment
  int i;                          // as selection
  for i = #x-1 to 0 step -1 {     // operator
    CNot(x[i],x[0::i] & e);
  }
}
```

Listing 8.4: Operator for incrementing quantum states in QCL defined as a conditional quantum function. Subroutine inc is defined using cond keyword and does not require the second argument of type quconst. Subroutine cinc provides equivalent implementation with explicit-declared enable register.

In the case of inc procedure, presented in Listing 8.4, the enable register is passed as an implicit argument. This argument is set by a quantum if-statement and transparently passed on to all suboperators. As a result, all suboperators have to be conditional. This is illustrated in the following example.

```
qcl> qureg q[4];qureg e[1];   // counting and control registers
qcl> H(q[3] & e);             // prepare test state
[5/32] 0.5 |0,0> + 0.5 |8,0> + 0.5 |0,1> + 0.5 |8,1>
qcl> cinc(q,e);               // conditional increment
```

```
[5/32] 0.5 |0,0> + 0.5 |8,0> + 0.5 |1,1> + 0.5 |9,1>
qcl> if e { inc(q); }          // equivalent to cinc(q,e)
[5/32] 0.5 |0,0> + 0.5 |8,0> + 0.5 |2,1> + 0.5 |10,1>
qcl> !cinc(q,e);               // conditional decrement
[5/32] 0.5 |0,0> + 0.5 |8,0> + 0.5 |1,1> + 0.5 |9,1>
qcl> if e { !inc(q); }         // equivalent to !cinc(q,e);
[5/32] 0.5 |0,0> + 0.5 |8,0> + 0.5 |0,1> + 0.5 |8,1>
```

QCL session in the above example presents computation involving conditional operations. QCL offers an interactive environment facilitating the testing of the quantum programs. After each comment, the interpreter reports the current state of the registers and thus it is possible to observe the results of the commands step by step.

Finally we should note that a conditional subroutine can be called outside a quantum if-statement. In such situation the enable register is empty and, as such, ignored. Subroutine call is in this case unconditional.

8.2   LANQ

The second quantum programming language based on the imperative paradigm is LanQ. As it has already been pointed out above, LanQ was developed to address the problems arising from the lack of elements supporting quantum communication.

Additionally, LanQ is the first quantum programming language with full operation semantics specified [155]. This allows for the formal reasoning about the correctness of programs written in LanQ and for the further development of the language. Semantics is also crucial for the optimization of the programs written in LanQ.

The programs written in LanQ can be executed and tested using an available interpreter [168]. The interpreter was developed as a part of a PhD thesis [155], but its development stopped in 2007.

8.2.1   BASIC ELEMENTS

The main feature of LanQ is the support for creating multipartite quantum protocols. LanQ, as well as cQPL presented in the next section, are built with quantum communication in mind. Thus, in contrast to QCL, they provide the features for facilitating the simulation of quantum communication.

The syntax of the LanQ programming language is very similar to the syntax of C programming language. In particular it supports:

- Classical data types: int and void.

- Conditional statements of the form

```
if ( cond ) {
  ...
} else {
```

```
    ...
}
```

- Looping with `while` keyword

```
while ( cond ) {
    ...
}
```

- User defined functions, for example

```
int fun( int i) {
    int res;
    ...
    return res;
}
```

8.2.2 PROCESS CREATION

LanQ is built around the concepts of process and interprocess communication, known for example from UNIX operating system. It provides the support for controlling quantum communication between many parties. The implementation of teleportation protocol presented in Listing 8.6 provides an example of LanQ features, which can be used to describe quantum communication.

Function `main()` in Listing 8.6 is responsible for controlling quantum computation. The execution of protocol is divided into the following steps:

1. Creation of the classical channel for communicating the results of measurement: `channel[int] c withends [c0,c1];`.

2. Creation of Bell state used as a quantum channel for teleporting a quantum state (`psiEPR aliasfor [psi1, psi2]`); this is accomplished by calling external function `createEPR()` creating an entangled state.

3. Instruction `fork` executes `alice()` function, which is used to be implement a sender; original process continues to run.

4. In the last step function `bob()` implementing a receiver is called.

8.2.3 COMMUNICATION

The communication between parties is supported by providing `send` and `recv` keywords. Communication is synchronous, i.e., `recv` delays the program execution until there is a value received from the channel and `send` delays the program run until the sent value is received.

```
void alice(channelEnd[int] c0, qbit auxTeleportState) {
  int i;
  qbit phi;
  // prepare state to be teleported
  phi = computeSomething();
  // Bell measurement
  i = measure (BellBasis, phi, auxTeleportState);
  send (c0, i);
}

void bob(channelEnd[int] c1, qbit stateToTeleportOn) {
  int i;
  i = recv(c1);
  // execute one of the Pauli gates according to the protocol
  if (i == 1) {
    Sigma_z(stateToTeleportOn);
  } else if (i == 2) {
    Sigma_x(stateToTeleportOn);
  } else if (i == 3) {
    Sigma_x(stateToTeleportOn);
    Sigma_z(stateToTeleportOn);
  }
  dump_q(stateToTeleportOn);
}
```

Listing 8.5: Modules used in the quantum teleportation program implemented in LanQ (see: Listing 8.6).

Processes can allocate *channels*. It should be stressed out that the notion of channels used in quantum programming is different from the one used in quantum mechanics. In quantum programming a channel refers to a variable shared between processes. In quantum mechanics a channel refers to *any quantum operation*.

Another feature used in quantum communication is variable aliasing. In the teleportation protocol presented in Listing 8.6 the syntax for variable aliasing

```
qbit psi1, psi2;
psiEPR aliasfor [psi1, psi2];
```

is used to create a quantum state shared among two parties.

```
void main() {
  channel[int] c withends [c0,c1];
  qbit psi1, psi2;
  psiEPR aliasfor [psi1, psi2];

  psiEPR = createEPR();

  c = new channel[int]();
  fork alice(c0, psi1);
  bob(c1, psi2);
}
```

Listing 8.6: Teleportation protocol implemented in LanQ [155]. Functions Sigma_x(), Sigma_y() and Sigma_z() are responsible for implementing Pauli matrices. Function createEPR() (not defined in the listing) creates maximally entangled state between parties—Alice and Bob. Quantum communication is possible by using the state, which is stored in a global variable psiEPR. Function computeSomething() (not defined in the listing) is responsible for preparing a state to be teleported by Alice.

8.2.4   TYPES

Types in LanQ are used to control the separation between classical and quantum computation. In particular they are used to prohibit copying of quantum registers. The language distinguishes two groups of variables [155, Chapter 5]:

- Duplicable or non-linear types for representing classical values, e.g., bit, int, boolean. The value of a duplicable type can be exactly copied.

- Non-duplicable or linear types for controlling quantum memory and quantum resources, e.g., qbit, qtrit channels and channel ends (see example in Listing 8.6). Types from this group do not allow for cloning [171].

One should note that quantum types defined in LanQ are mainly used to check validity of the program before its run. However, such types do not help to define abstract operations. As a result, even simple arithmetic operations have to be implemented using elementary quantum gates, e.g., using quantum circuits introduced in [172].

8.3   SUMMARY

We have presented only the basic elements of two existing imperative quantum programming languages which, thanks to the availability of the interpreters, can be used to execute and test quantum programs.

As one can see most of the construction provided in theses languages is very similar to the elements of the quantum pseudocode, introduced in Chapter 5. Both QCL and LanQ provide a syntax familiar for any user with a basic knowledge of any modern programming language.

CHAPTER 9

Functional Quantum Programming

Classical programmers have found many methods for dealing with the growing complexity of created programs. One of them is the functional approach for creating programming languages.

Classical functional programming languages have many features which allow us to clearly express algorithms [19, 173]. In particular they allow for writing better-modularized programs than in the case of imperative programming languages. Functional programming languages encourage us to write programs which are well structured and easier to read and to maintain than in the case of programs written in imperative languages. These features are important for software developers since thanks to them functional languages allow us to debug programs more easily and reuse software components. Both aspects are crucial especially in large and complex software projects.

The program in a functional programming language is written as a function, which is defined in terms of other functions. Classical functional programming languages contain no assignment statements, and this allows us to eliminate side-effects.[1] It means that the function call can have no effect other than to compute its result [173]. In particular it cannot change the value of a global variable.

As we have already pointed out, the lack of progress in creating new quantum algorithms is caused by the problems with operating on complex quantum structures—multi-qubit quantum states and quantum gates operating on them. The problem with understanding the quantum computational process motivated the research in the area of quantum functional programming languages. Quantum functional programming attempts to merge the concepts known from classical functional programming practice with the elements used in quantum information processing. As we will see this is not always possible due to the elementary properties of the quantum world.

During the last few years few quantum programming languages based on functional programming paradigm have been proposed [174]. The first attempts to define a functional quantum programming language were made by using quantum lambda calculus [157], which was based on lambda calculus. One can note that most of the research in this area is focused on the formal properties of quantum programming languages and the functional paradigm is used mainly as a method for studying these properties.

We begin our review by discussing some tools developed using classical functional programming languages. These tools, usually developed in the form of packages of functions, aim to bring the

---

[1]This is true in *pure functional* programming languages like Haskell.

use of functional programming languages in order to widen the understanding of quantum mechanics. The main obstacle for using these tools, however, is that they are heavily based on the concepts introduced in quantum functional programming.

We aim to introduce the concepts used in functional quantum programming with the help of working examples. For this reason in this chapter we focus on two functional quantum programming languages—cQLP and QML[2]—which have working interpreters.

Here, however, we aim to continue to use a pragmatic approach and we focus on high-level quantum programming languages rather than on the simulation. The most important feature of these languages, in comparison to the modelling tools based on classical languages, is the use of relatively simple syntax developed for the purpose of expressing elements used in quantum computation. For this reason programming languages—like cQPL and QML discussed below—represent the best of two worlds. They allow us to easily express the quantum computational process and, at the same time, they bring the power of functional programming into the quantum domain.

9.1   FUNCTIONAL MODELLING OF QUANTUM COMPUTATION

Advocates of functional programming often argue that the languages of this type allow us to express the complex structures in a more readable manner, compared to the imperative languages, and that this facilities the reasoning about the written programs [175]. It is not a surprise that a considerable research effort has been devoted to the problem of research on modelling quantum computation using functional programming.

As Haskell is *de facto* standard in the community of researchers working on functional programming, some of the research in this area was conducted using this language for the implementation purposes.

The usefulness of functional programming for the purpose of modelling quantum programming was for the first time considered by Bird and Mu [176]. In this work a monadic style of quantum programming is proposed and the quantum programming is considered a special case of the non-deterministic programming.

Skibiński developed one of the first simulators of quantum computing written in functional programming language [177]. The simulator is currently available from the official Haskell repository.

One of the first attempts at using functional programming for the purpose of quantum computing was based on the extension of the λ-calculus. It was due to van Tonder [157, 178], who defined a quantum λ-calculus with type system based on linear logic [179]. However, quantum λ-calculus deals with the pure quantum computation only. As such it does not introduce the measurement required for the purpose of modelling classically controlled quantum computation. The simulator of the quantum λ-calculus was also developed [180] using Scheme programming languages.

---

[2]Not to be confused with a XML-based language used in the description of quantum circuits introduced in Chapter 4.

Among the subsequent attempts at using functional programming language for this purpose one can point out the papers by Sabry [181] and Karczmarczuk [158].

In [181] Sabry developed a framework for simulating quantum computation. He also discussed the problems arising in the use of functional programming for this purpose.

Karczmarczuk [158] proposed a functional framework for quantum computing formulated in Haskell programming language. He developed an abstract geometric framework which can be used to simulate quantum mechanical objects in terms of functional programming.

The reader interested in the recent development in this area is advised to consult the on-line list of [162]. It provides a community updated catalogue of software used to model quantum computation, including the tools based on the functional paradigm.

9.2   CQPL

We start with a brief description of cQPL language developed by Maurer [148]. This language is directly related to QPL (Quantum Programming Language) proposed by Selinger [159]. Although very influential for the further developments in the theory of quantum programming, QPL has never been implemented. On the other hand, the cQPL compiler was implemented [148] and is available from its author upon request. The compiler was written in OCaml and uses the libqc simulation library developed by Ömer [136, 137].

In [159] Quantum Programming Language (QPL) was described and in [148] its extension useful for the modelling of quantum communication was proposed. This extended language was named cQPL—communication capable QPL. Since cQPL compiler is also QPL compiler, we will describe cQPL only.

The compiler for cQPL language described in [148] is built on the libqc simulation library used in the QCL interpreter. As a result, cQPL provides some features known from QCL.

9.2.1   CLASSICAL ELEMENTS

Classical elements of cQPL are very similar to classical elements implemented in imperative programming languages. The syntax resembles that of classical programming languages based on C programming language.

In particular cQPL provides conditional structures using if ... then ... else block and loops are introduced with while keyword.

To improve modularity cQPL provides the support for procedures. The can be introduced using proc keyword. As expected, procedures can accept classical as well as quantum data as the input parameters.

```
proc test: a:int, q:qbit {
   ...
}
```

```
new int loop := 10;
while (loop > 5) do {
    print loop;
    loop := loop - 1;
};
if (loop = 3) then {
    print "loop is equal 3";
} else {
    print "loop is not equal 3";
};
```

Listing 9.1: Classical elements of cQPL. Variables are declared using new keyword. Classical control structures include while loop and if/then/else conditional statement.

Procedure call has to know the number of parameters returned by the procedure. If, for example, procedure test is defined as above, it is possible to gather the calculated results

```
new int a1 = 0;
new int cv = 0;
new int qv = 0;
(a1) := call test(cv, qv);
```

or ignore them

```
call test(cv, qv);
```

In the first case the procedure returns the values of input variables calculated at the end of its execution.

Classical variables are passed by value i.e., their value is copied. This is impossible for quantum variables, since a quantum state cannot be cloned [171]. Thus, it is also impossible to assign the value of a quantum variable calculated by the procedure.

Note that no cloning theorem requires quantum variables to be *global*. This shows that in the quantum case it is impossible to avoid some effects known from imperative programming and typically not present in functional programming languages.

Global quantum variables are used in Listing 9.3 to create a maximally entangled state in a teleportation protocol. Procedure createEPR(epr1, epr2) operates on two quantum variables (subsystems) and produces a Bell state.

9.2.2   QUANTUM ELEMENTS

Quantum Registers

Quantum memory can be accessed in cQPL using the variables of type `qbit` or `qint`. Basic operations on quantum registers are presented in Listing 9.2. In particular, the execution of a quantum gates is performed by using `*=` operator.

```
new qbit q1 := 0;
new qbit q2 := 1;
// execute CNOT gate on both qubits
q1, q2 *= CNot;
// execute phase gate on the first qubit
q1 *= Phase 0.5;
```

Listing 9.2: State initialization and basic gates in cQPL. Data type `qbit` represents a single qubit.

It should be pointed out that `qint` data type provides only a short-cut for accessing the table of qubits.

Quantum Gates

Only a few elementary quantum gates are built into the language:

- Single qubit gates `H`, `Phase` and `NOT` implementing elementary one-qubits gates listed in Table 4.2 in Chapter 4,

- `CNot` operator implementing controlled negation,

- `FT(n)` operator for n-qubit quantum Fourier transform.

This set of operations allows us to simulate an arbitrary quantum computation. Besides, it is possible to define new gates by specifying their matrix elements directly.

Measurement

Measurement is performed using `measure/then` keywords and `print` command allows us to display the value of a variable.

```
measure a then {
  print "a is |0>";
} else {
  print "a is |1>";
};
```

Similarly to QCL, it is also possible to inspect the value of a state vector using `dump` command.

9.2.3 QUANTUM COMMUNICATION

The main feature of cQPL is its ability to build and test quantum communication protocols easily. Communicating parties are described using *modules*. Analogous to LanQ, cQPL introduces channels which can be used to send quantum data.

Once again we stress that the notion of channels used in cQPL and LanQ is different from that used in the quantum theory. In quantum mechanics channels, sometimes referred to as operations, are used to describe allowed physical transformations, while in quantum programming they are used to describe communication links.

Communicating parties are described by modules, introduced using `module` keyword. Modules can exchange quantum data (states). This process is accomplished using `send` and `receive` keywords.

To compare cQPL and LanQ one can use the implementation of the teleportation protocol. The implementation of teleportation protocol in cQPL is presented in Listing 9.3, while the implementation in LanQ is provided in Listing 8.6.

```
module Alice {
    proc createEPR: a:qbit, b:qbit {
       a *= H;
       b,a *= CNot;   /* b: Control, a: Target */
    } in {
      new qbit teleport := 0;
      new qbit epr1 := 0;
      new qbit epr2 := 0;

      call createEPR(epr1, epr2);
      send epr2 to Bob;

      /* teleport: Control, epr1: Target  (see: Figure 4.4)  */
      teleport, epr1 *= CNot;

      new bit m1 := 0;
      new bit m2 := 0;
      m1 := measure teleport;
      m2 := measure epr1;

      /* Transmit the classical measurement results to Bob */
      send m1, m2 to Bob;
};

module Bob {
```

```
   receive q:qbit from Alice;
   receive m1:bit, m2:bit from Bob;

   if (m1 = 1) then { q *= [[ 0,1,1,0 ]];
/* Apply sigma_x */ };

   if (m2 = 1) then { q *= [[ 1,0,0,-1 ]];
/* Apply sigma_z */};

   /* The state is now teleported */
   dump q;
};
```

Listing 9.3: Teleportation protocol implemented in cQPL (from [148]). Two parties—Alice and Bob—are described by modules. Modules in cQPL are introduced using module keyword and can exchange quantum data using send/receive structure.

As we mentioned in Chapter 1, programming languages are in many cases based on the mixed paradigm. In the case of cQPL, the modules resemble to some extent the objects used in object-oriented languages.

9.3   QML

Another quantum programming language following a functional paradigm is QML developed by Altenkirch and Grattage [132, 160]. The QML compiler was described in [133] and can be downloaded from the project web page [182].

The name suggests that QML was designed as a quantum version of ML language [58]. The language, however, is implemented in Haskell (see e.g. [59, 60]) and follows some syntactic conventions used in Haskell.

The QML compiler requires GHC (Glasgow Haskell Compiler) [183] in version 6 in order to run QML programs. In order to run a program written in QML one needs to load the definitions in qml.hs into the interactive environment ghci and use one of the functions described in Table 9.1.

9.3.1   PROGRAM STRUCTURE

A program written in QML consists of a sequence of function definitions. Each definition is of the form

```
funName (var1,type1) ... (varN,typeN) |- funBody :: retType
```

Table 9.1: Possible methods of evaluation of QML programs [133].	
Function	**Evaluation method**
runM	Unitary matrix representing a reversible part of the program.
runI	Isometry providing a full description of the program for terms that produce no garbage.
runS	Superoperator initializing the heap and tracing-out the garbage.

For example, the classical not function (Cnot) is defined as

```
Cnot (q,qb) |- if q then qtrue
                    else qfalse :: qb
```

Using the same syntax the user can define constants, which in QML are equivalent to functions. For example, to use a constant representing a superposition $\frac{|\rangle}{0} + |1\rangle \sqrt{2}$ one can declare it as

```
-- one-qubit superposition
Qsup |- hF*qtrue + hF*qfalse :: qb
```

Here the term hF is defined as $\frac{1}{\sqrt{2}}$. The $*$ operation allows us to associate the probability amplitude with a term.

The state of compound systems can be represented using the () operation. For example, the constant representing the EPR pair (i.e., one of the Bell states) is defined as

```
Epr |- hF * (qtrue,qtrue) + hF * (qfalse,qfalse) :: qb*qb;
```

One should note that in the above example the resulting type is described as qb*qb and here $*$ operator is used to describe a type of two-qubit state.

9.3.2 SUBROUTINES

As in any functional programming language, programs written in QML are composed of small functions. This makes the written code more readable and easier to maintain.

Subroutines in QML can operate on an arbitrary number of arguments. A subroutine is introduced by the following syntax

```
ProcName (arg) -|    code
                     code
```

One should note that procedure names have to start with the upper case letter. Moreover, similarly to Haskell, the indentation is important and denotes the continuation of the code block.

```
Cnot (b,qb) |- if b then qfalse else qtrue :: qb;


CNot (s,qb) (b,qb) |- if s then Cnot (b) else b :: qb;

-- classically-controlled quantum Not
CQnot (s,qb) (t,qb) |- if s then Qnot (t) else t :: qb;
```

The quantum CNot operation can be defined in terms of the above by using a quantum conditional operation

```
QCNot (s,qb) (t,qb) |- ifo s then (qtrue,Qnot (t))
                              else (qfalse,t) :: qb * qb;
```

where Qnot is defined as

```
Qnot (b,qb) |- ifo b then qfalse else qtrue :: qb;
```

The program for the quantum teleportation is presented in Listing 9.7.

```
-- The constant EPR pair
Epr |- hF * (qtrue,qtrue) + hF * (qfalse,qfalse) :: qb*qb;
```

Listing 9.4: Declaration of the maximally entangled state in QML.

```
-- The correction operations
Uol (x,qb) |- ifo x then  qfalse else qtrue  :: qb;
Ulo (x,qb) |- ifo x then -qtrue  else qfalse :: qb;
Ull (x,qb) |- ifo x then -qfalse else qtrue  :: qb;

-- The "unitary correction"
U (q,qb) (xy,qb*qb) |- let (x,y) = xy
                       in if x
                          then (if y then Ull (q) else Ulo (q))
                          else (if y then Uol (q) else q) :: qb;
```

Listing 9.5: Correction step implemented using quantum conditions.

```
-- The measurement operator, using "if"
Meas (x,qb) |- if x then qtrue else qfalse :: qb;

-- The Bell-measurement operation
Bmeas (x,qb) (y,qb) |- let (xa,ya) = CNot (x,y)
```

```
                          in  (Meas (Had (xa)),Meas (ya)) :: qb*qb;

Bnmeas (x,qb) (y,qb) |- let (xa,ya) = CNot (x,y)
                          in  (Had (xa),ya) :: qb*qb;
```

Listing 9.6: Function implementing single-qubit measurement and Bell measurement in QML.

```
-- The main procedure for the teleportation protocol
Tele (a,qb) |- let (b,c) = Epr ()
            in  let f = Bnmeas (a,b)
            in U (c,f) :: qb;
```

Listing 9.7: Quantum teleportation in QML [133]. The procedure consists of three steps. First the shared entangled state is created. Next, the appropriate measurement is carried out and the result of this measurement is used in the last step. Appropriate functions are defined in Listings 9.4, 9.5, and 9.6.

In order to run the program from Listing 9.7, one should type

```
ghci> runTC "teleport.qml" "Tele"
```

for the typed circuit, or

```
ghci> runI "teleport.qml" "Tele"
```

for the isometry [133].

9.4 SUMMARY

In this chapter we have discussed the research aimed at using functional programming methods in quantum information theory. We presented two quantum programming languages designed to bring functional style of programming into the quantum domain. We have also briefly presented the research aimed at using existing functional programming languages for the purpose of simulating quantum systems.

The potential for using functional programming languages in scientific computing and, in particular, in modelling quantum mechanical objects was for the first time discussed by Karczmarczuk [184]. A more pragmatic approach was presented in [185], where a symbolic simulation of quantum algorithms in *Mathematica* is considered. Nevertheless, the functional methods are still inferior in popularity, at least if one compares the number of software tools based on functional paradigm with the number of tools using the imperative style for modelling quantum computation [162].

We have mentioned that the main motivation for considering quantum functional programming lies in the ability to use functional methods for the purpose of reasoning about programs. Among the recent tools developed for this purpose one can point out Quantomatic [186]. This tool provides an abstract and symbolic way to represent and simulate quantum information processing based on graph-based formalisms for computation [187].

One of the features connected with functional programming languages is garbage collection [188], proposed for the first time to solve problems in Lisp. The ability to discard unused registers can have a great impact on the efficiency of the physical realization of the high-level program. However, this functionality can be introduced on different levels of translation in the architecture described in Chapter 6.

CHAPTER 10

Outlook

In only 100 pages it is impossible to cover so broad and lively developed areas of knowledge. Classical computational models have been developed for more than fifty years now. On the other hand the research in quantum information and quantum computation theory, in the last two decades of the last century, brought us an explosion of new results related to the models of computation, as well as the laws of physics. For this reason we believe that this book should be seen more as a guided tour of the most important results in these areas rather than a comprehensive coverage of the presented topics.

The main goal of this book was to acquaint the reader with quantum programming languages and computational models used in quantum information theory. We have described the quantum Turing machine, quantum circuits, and QRAM models of quantum computation. We have also presented four quantum programming languages—QCL, LanQ, cQPL, and QML. The common feature of these languages is that there exists a working interpreter allowing us to actually run a program written in one of these languages. This allows the reader to experiment with the presented examples and gain some insight into the laws of quantum mechanics by actually writing programs for quantum computers.

The presented quantum programming languages have many advantages when compared to the circuit model. Firstly one can note that the syntax of presented languages resembles the syntax of popular classical programming languages like C [56] or Java [62]. As such, it can be easily mastered by programmers familiar with classical languages and, moreover, the description of quantum algorithms in quantum programming languages is better suited for people unfamiliar with the notion used in quantum mechanics. The second advantage is their ability to use classical control structures for controlling the execution of quantum operators. The next advantage of quantum programming languages is the ability to reason about the correctness of the written programs. Additionally, some of the presented languages provide the syntax for clear description of communication protocols.

The above features make the quantum programming languages a great tool for studying quantum information theory and quantum mechanics.

The important problem is the expressiveness of the languages. One can note that the languages presented in this book provide a very similar set of basic quantum gates and allow us to operate only on the arrays of qubits. Most of the gates provided by these languages correspond to the basic quantum gates. Thus, one can conclude that the presented languages have the ability to express quantum algorithms similar to the abilities of a quantum circuit model.

In our opinion the main disadvantage of the described languages is the lack of quantum data types. The types defined in described languages are used mainly for two purposes: to optimize the memory management in QCL and to avoid compile-time errors caused by copying quantum registers in cQPL and LanQ. Both reasons are important from the simulations point of view, since they facilitate writing of correct and optimized quantum programs. However, these features do not provide a mechanism for developing new quantum algorithms or protocols. That is why we believe that there is still much to be discovered in the area of quantum programming.

Bibliography

[1] J.A. Miszczak. *Probabilistic aspects of quantum programming*. PhD thesis, Institute of Theoretical and Applied Informatics, Polish Academy of Sciences, March 2008. Document available on-line at `http://www.iitis.pl/~miszczak/papers/`. Cited on page(s) xv

[2] P. Gawron and J.A. Miszczak. Numerical simulations of mixed state quantum computation. *Int. J. Quant. Inf.*, 2:195–199, 2005. DOI: 10.1142/S0219749905000748 Cited on page(s) xv

[3] J.A. Miszczak. Description and visualisation of quantum circuits with XML. *Theoretical and Applied Informatics (formerly Archiwum Informatyki Teoretycznej i Stosowanej)*, 17(4):265–272, 2005. Cited on page(s) xv, 61

[4] P. Gawron, J. Klamka, J.A. Miszczak, and R. Winiarczyk. Extending scientic computing system with structural quantum programming capabilities. *Bull. Pol. Acad. Sci.-Tech. Sci.*, 58(1):77–88, 2010. DOI: 10.2478/v10175-010-0008-4 Cited on page(s) xv, 68, 70

[5] J.A. Miszczak. Models of quantum computation and quantum programming languages. *Bull. Pol. Acad. Sci.-Tech. Sci.*, 59(3):305–324, 2011. DOI: 10.2478/v10175-011-0039-5 Cited on page(s) xv

[6] M. A. Nielsen and I. L. Chuang. *Quantum Computation and Quantum Information*. Cambridge University Press, Cambridge, U.K., 10th anniversary edition, 2010. Cited on page(s) 1, 14, 15, 33, 37, 55, 66

[7] M. Lanzagorta and J. Uhlmann. *Quantum Computer Science*, volume 2 of *Synthesis Lectures on Quantum Computing*. Morgan & Claypool Publishers, 2009. DOI: 10.2200/S00159ED1V01Y200810QMC002 Cited on page(s) 1

[8] H.M. Wiseman and G.J Milburn. *Quantum Measurement and Control*. Cambridge University Press, Cambridge, U.K, 2010. Cited on page(s) 1

[9] T.S. Metodi, A.I. Faruque, and F.T. Chong. *Quantum Computing for Computer Architects*, volume 13 of *Synthesis Lectures on Computer Architecture*. Morgan & Claypool Publishers, second edition, 2011. DOI: 10.2200/S00066ED1V01Y200610CAC001 Cited on page(s) 1, 51, 63

[10] M. Fernández. *Models of Computation: An Introduction to Computability Theory*. Springer Verlag, London, UK, 2008. Cited on page(s) 1

[11] C. H. Papadimitriou. *Computational complexity*. Addison-Wesley Publishing Company, 1994. Cited on page(s) 1, 2, 7, 8, 9, 11, 17, 35, 46

[12] S. Arora and B. Barak. *Computational Complexity. A Modern Approach*. Cambridge University Press, Cambridge , U.K., 2009. Cited on page(s) 1, 2, 7

[13] C.E. Shannon. A symbolic analysis of relay and switching circuits. Master's thesis, Massachusetts Institute of Technology, 1937. Cited on page(s) 1, 33

[14] H. Vollmer. *Introduction to Circuit Complexity*. Springer-Verlag, Berlin/Heidelberg, Germany, 1999. Cited on page(s) 1

[15] S. A. Cook and R. A. Reckhow. Time-bounded random access machines. In *Proceedings of the forth Annual ACM Symposium on Theory of Computing*, pages 73–80, 1973. DOI: 10.1145/800152.804898 Cited on page(s) 1, 45, 46

[16] J. C. Shepherdson and H. E. Strugis. Computability of recursive functions. *J. ACM*, 10(2):217–255, April 1963. DOI: 10.1145/321160.321170 Cited on page(s) 1, 45

[17] A. Church. An unsolvable problem of elementary number theory. *American Journal of Mathematics*, 58:345–363, 1936. DOI: 10.2307/2371045 Cited on page(s) 2

[18] H. Abelson, G. J. Sussman, and J. Sussman. *Structure and Interpretation of Computer Programs*. MIT Press, Cambridge, Massachusetts, U.S.A., 2nd edition, 1996. Cited on page(s) 2

[19] J. C. Mitchell. *Concepts in programming languages*. Cambridge University Press, Cambridge, U.K., 2003. Cited on page(s) 2, 5, 83

[20] D. Deutsch. Quantum theory, the Church-Turing principle and the universal quantum computer. *Proc. R. Soc. Lond. A*, 400:97, 1985. DOI: 10.1098/rspa.1985.0070 Cited on page(s) 2, 12, 75

[21] C. Bohm. On a family of Turing machines and the related programming language. *ICC Bull.*, 3:187–194, 1964. Cited on page(s) 3, 46

[22] G. E. Moore. Cramming more components onto integrated circuits. *Electronics*, 38(8):114–117, 1965. DOI: 10.1109/JPROC.1998.658762 Cited on page(s) 3

[23] Moores law: Intel microprocessor transistor count chart. Data available on-line at http://www.intel.com/about/companyinfo/museum/exhibits/moore.htm. Cited on page(s) 4

[24] R. P. Feynman. Simulating physics with computers. *Int. J. Theor. Phys.*, 21(467):467, 1982. DOI: 10.1007/BF02650179 Cited on page(s) 3

[25] C. H. Bennett and G. Brassard. Quantum cryptography: public key distribution and coin tossing. In *Proceedings of the IEEE International Conference on Computers, Systems, and Signal Processing, Bangalore, India*, pages 175–179, 1984. DOI: 10.1016/j.tmaid.2008.06.006 Cited on page(s) 3, 66

[26] A. Ekert. Quantum cryptography based on Bell's theorem. *Phys. Rev. Lett.*, 67:661–663, 1991. DOI: 10.1103/PhysRevLett.67.661 Cited on page(s) 3

[27] A. Einstein, B. Podolsky, and N. Rosen. Can quantum-mechanical description of physical reality be considered complete? *Phys. Rev.*, 47(10):777–780, 1935. DOI: 10.1103/PhysRev.47.777 Cited on page(s) 3

[28] J. Bouda. *Encryption of Quantum Information and Quantum Cryptographic Protocols.* PhD thesis, Faculty of Informatics, Masaryk University, 2004. Cited on page(s) 3

[29] R. Ursin, F. Tiefenbacher, T. Schmitt-Manderbach, H. Weier, T. Scheidl, M. Lindenthal, B. Blauensteiner, T. Jennewein, J. Perdigues, P. Trojek, B. Ömer, M. Fürst, M. Meyenburg, J. Rarity, Z. Sodnik, C. Barbieri, H. Weinfurter, and A. Zeilinger. Entanglement-based quantum communication over 144 km. *Nat. Phys.*, 3:481–486, 2007. DOI: 10.1038/nphys629 Cited on page(s) 4

[30] M. Peev, C. Pacher, R. Alléaume, C. Barreiro, J. Bouda, W. Boxleitner, T. Debuisschert, E. Diamanti, M. Dianati, J. F. Dynes, S. Fasel, S. Fossier, M. Fürst, J. D. Gautier, O. Gay, N. Gisin, P. Grangier, A. Happe, Y. Hasani, M. Hentschel, H. Hübel, G. Humer, T. Länger, M. Legré, R. Lieger, J. Lodewyck, T. Lorünser, N. Lütkenhaus, A. Marhold, T. Matyus, O. Maurhart, L. Monat, S. Nauerth, J. B. Page, A. Poppe, E. Querasser, G. Ribordy, S. Robyr, L. Salvail, A. W. Sharpe, A. J. Shields, D. Stucki, M. Suda, C. Tamas, T. Themel, R. T. Thew, Y. Thoma, A. Treiber, P. Trinkler, R. Tualle-Brouri, F. Vannel, N. Walenta, H. Weier, H. Weinfurter, I. Wimberger, Z. L. Yuan, H. Zbinden, and A. Zeilinger. The SECOQC quantum key distribution network in Vienna. *New J. Phys.*, 11(7):075001, 2009. DOI: 10.1088/1367-2630/11/7/075001 Cited on page(s) 4

[31] D.J. Rogers. *Broadband Quantum Cryptography.* Synthesis Lectures on Computer Architecture. Morgan & Claypol Publising, 2010. DOI: 10.2200/S00265ED1V01Y201004QMC003 Cited on page(s) 4

[32] P. W. Shor. Algorithms for quantum computation: Discrete logarithms and factoring. In *Proceedings of the 35th Annual Symposium on Foundations of Computer Science*, pages 124–134. IEEE Computer Society Press, 1994. DOI: 10.1109/SFCS.1994.365700 Cited on page(s) 4, 5, 66

[33] P. Shor. Polynomial-time algorithms for prime factorization and discrete logarithms on a quantum computers. *SIAM J. Computing*, 26:1484–1509, 1997. DOI: 10.1137/S0097539795293172 Cited on page(s) 4, 5

[34] L. K. Grover. Quantum mechanics helps in searching for a needle in a haystack. *Phys. Rev. Lett.*, 79:325–328, 1997. DOI: 10.1103/PhysRevLett.79.325 Cited on page(s) 4, 66, 73, 74

[35] J. Kempe. Quantum random walks: An introductory overview. *Contemp. Phys.*, 44(4):307–327, 2003. DOI: 10.1080/00107151031000110776 Cited on page(s) 4

[36] J. Košík. Two models of quantum random walk. *Cent. Eur. J. Phys.*, 4:556–573, 2003. DOI: 10.2478/BF02475903 Cited on page(s) 4

[37] J. Eisert, M. Wilkens, and M. Lewenstein. Quantum games and quantum strategies. *Phys. Rev. Lett.*, 83:3077–3080, 1999. DOI: 10.1103/PhysRevLett.83.3077 Cited on page(s) 4

[38] D.A. Meyer. *AMS Contemporary Mathematics: Quantum Computation and Quantum Information Science*, volume 305, chapter Quantum games and quantum algorithms. American Mathematical Society, Providence, Rhode Island, U.S.A. Cited on page(s) 4

[39] S.E. Venegas-Andraca. *Quantum Walks for Computer Scientists*, volume 1 of *Synthesis Lectures on Quantum Computing*. Morgan & Claypool Publishers, 2008. DOI: 10.2200/S00144ED1V01Y200808QMC001 Cited on page(s) 4

[40] S.E. Venegas-Andraca. Quantum walks: a comprehensive review. *Arxiv preprint arXiv:1201.4780*, 2012. Cited on page(s) 4

[41] A. Ambainis. Quantum walk algorithm for element distinctness. *SIAM Journal on Computing*, 37:210–239, 2007. DOI: 10.1137/S0097539705447311 Cited on page(s) 4

[42] A. M. Childs and J. M. Eisenberg. Quantum algorithms for subset finding. *Quantum. Inf. Comput.*, 5:593, 2005. Cited on page(s) 4

[43] A. Ambainis, J. Kempe, and A. Rivosh. Coins make quantum walks faster. In *Proceedings of the sixteenth annual ACM-SIAM symposium on Discrete algorithms*, pages 1099–1108. Society for Industrial and Applied Mathematics, 2005. Cited on page(s) 4

[44] F. Magniez, M. Santha, and M. Szegedy. Quantum algorithms for the triangle problem. In *Proceedings of the sixteenth annual ACM-SIAM symposium on Discrete algorithms*, pages 1109–1117. Society for Industrial and Applied Mathematics, 2005. DOI: 10.1137/050643684 Cited on page(s) 4

[45] H. Buhrman and R. Špalek. Quantum verification of matrix products. In *Proceedings of the seventeenth annual ACM-SIAM symposium on Discrete algorithm*, pages 880–889. ACM, 2006. DOI: 10.1145/1109557.1109654 Cited on page(s) 4

[46] A. Ambainis. Quantum walks and their algorithmic applications. *Int. J. Quant. Inf.*, 1:507–518, 2003. DOI: 10.1142/S0219749903000383 Cited on page(s) 4

[47] P. W. Shor. Progress in quantum algorithms. *Quantum Information Processing*, 3(1-5), 2004. DOI: 10.1007/s11128-004-3878-2 Cited on page(s) 5, 66

[48] S. J. Lomonaco and L.K Kauffman. Search for new quantum algorithms. Technical Report F30602-01-2-0522, Defense Advanced Research Projects Agency (DARPA) and Air Force Research Laboratory, Air Force Materiel Command, USAF, 2005. DOI: 10.1145/992287.992296 Cited on page(s) 5

[49] E. Bernstein and U. Vazirani. Quantum complexity theory. *SIAM J. Computing*, 26(5):1411–1473, 1997. DOI: 10.1137/S0097539796300921 Cited on page(s) 5, 13, 16, 17, 66

[50] L. Fortnow. One complexity theorist's view of quantum computing. *Theor. Comput. Sci.*, 292(3):597–610, 2003. DOI: 10.1016/S0304-3975(01)00377-2 Cited on page(s) 5, 16, 17

[51] S. Jordan. Quantum algorithm ZOO. On-line catalog of quantum algorithms offering speedup over the fastest classical algorithms published at `http://math.nist.gov/quantum/zoo/`. Cited on page(s) 5

[52] T.J. Bergin. A history of the history of programming languages. *Commun. ACM*, 50:69–74, 2007. DOI: 10.1145/1230819.1230841 Cited on page(s) 5

[53] M. Fernández. *Programming Languages and Operational Semantics*, volume 1 of *Texts in Computing*. King's College Publications, London, U.K., 2004. Cited on page(s) 5

[54] IBM. Preliminary report. specifications for the IBM mathematical FORmula TRANslating system. Technical report, IBM Corporation, 1954. Cited on page(s) 6

[55] N. Wirth. The programming language Pascal. *Acta Inform.*, 1(1):35–63, 1971. DOI: 10.1007/BF00264291 Cited on page(s) 6

[56] B. W. Kernighan and D. M. Ritchie. *C Programming Language*. Prentice Hall, Upper Saddle River, N.J., U.S.A., 2nd edition, 1988. Cited on page(s) 6, 72, 95

[57] J. McCarthy, P. Abrahams, D. Edwardsw, T. Hart, and M. Levin. *LISP 1.5 Programmer's Manual*. MIT Press, Cambridge, MA, 1965. Cited on page(s) 6

[58] R. Milner, M. Tofte, R. Harper, and D. MacQueen. *The Definition of Standard ML, Revised Edition*. The MIT Press, 1997. Cited on page(s) 6, 89

[59] G. Hutton. *Programming in Haskell*. Cambridge University Press, Cambridge, U.K., 2007. Cited on page(s) 6, 89

[60] M. Lipovača. *Learn You a Haskell for Great Good!* No Starch Press, 2011. Book avaible on-line at `http://learnyouahaskell.com/`. Cited on page(s) 6, 89

[61] A. Colmerauer and P. Roussel. The birth of Prolog. *SIGPLAN Not.*, 28:37–52, March 1993. DOI: 10.1145/155360.155362 Cited on page(s) 6

[62] J. Gosling, B. Joy, G. Steele, and G. Bracha. *Java(TM) Language Specification*. Prentice Hall PTR, 3rd edition, 2005. Cited on page(s) 6, 95

[63] M. Pilgrim. *Dive Into Python 3*. Apress, 2009. Book avaible on-line at http://www.diveintopython3.net/. DOI: 10.1007/978-1-4302-2416-7 Cited on page(s) 6

[64] J.E. Hopcroft, R. Motwani, and J.D. Ullman. *Introduction to Automata Theory, Languages, and Computation*. Addison Wesley, 2nd edition, 2000. Cited on page(s) 8, 19

[65] S. Aaronson and G. Kuperberg. Complexity ZOO. On-line aencyclopedia avaible at http://qwiki.stanford.edu/wiki/Complexity_Zoo. Cited on page(s) 12, 17

[66] A. Yao. Quantum circuit complexity. In *Proceedings of the 34th IEEE Symposium on Foundations of Computer Science*, pages 352–360. IEEE Computer Society Press, 1993. DOI: 10.1109/SFCS.1993.366852 Cited on page(s) 12, 17

[67] H. Nishimura and M. Ozawa. Computational complexity of uniform quantum circuit families and quantum turing machines. *Theor. Comput. Sci.*, 276:147–181, 2002. DOI: 10.1016/S0304-3975(01)00111-6 Cited on page(s) 12

[68] S. Iriyama, M. Ohya, and I. Volovich. Generalized quantum Turing machine and its application to the SAT chaos algorithm. In L. Accardi, M. Ohya, and N. Watanabe, editors, *Quantum Information and Computing*, volume 19 of *QP-PQ: Quantum Probability and White Noise Analysis*, 2007. Cited on page(s) 14

[69] S. Perdrix and P. Jorrand. Classically-controlled quantum computation. *Math. Struct. Comp. Sci.*, 16:601–620, 2006. DOI: 10.1017/S096012950600538X Cited on page(s) 15

[70] U. Vazirani. A survey of quantum complexity theory. *Proc. Sympos. Appl. Math.*, 58, 2002. DOI: 10.1137/S0097539796300921 Cited on page(s) 16

[71] D.C Kozen. *Theory of Computation: Classical and Contemporary Approaches*. Springer, 2006. Cited on page(s) 16

[72] M. R. Garey and D. S. Johnson. *Computers and Intractability: A Guide to the Theory of NP-Completeness*. W. H. Freeman, 1979. Cited on page(s) 17

[73] S. Aaronson. Quantum computing, postselection, and probabilistic polynomial-time. *Proc. R. Soc. A*, 461:3473–3482, 2005. DOI: 10.1098/rspa.2005.1546 Cited on page(s) 17

[74] G.S. Boolos, J.P. Burgess, and R.C. Jeffrey. *Computability and Logic*. Cambridge University Press, Cambridge, U.K., 5th edition, 2007. Cited on page(s) 18, 51

[75] G. H. Mealy. A method to synthesizing sequential circuits. *Bell Systems Technical Journal*, pages 1045–1079, 1955. Cited on page(s) 19, 20

[76] E.F Moore. Gedanken-experiments on sequential machines. In C.E. Shannon and J. Mc-Carthy, editors, *Automata Studies*, volume 34, pages 129–153. Princeton University Press, Princeton, N.J., 1965. Cited on page(s) 19, 20

[77] S.C. Kleen. Representation of events in nerve nets and finite automata. In S.C. Shannon and J. McCarthy, editors, *Automata Studies*, volume 34, pages 3–41. Princeton University Press, NJ, 1956. Cited on page(s) 19

[78] M.O. Rabin and D. Scott. Finite automata and their decision problems. *IBM J. Res. Dev.*, 3(2):114–125, 1959. DOI: 10.1147/rd.32.0114 Cited on page(s) 21, 23

[79] J.C. Shepherdson. The reduction of two-way automata to one-way automata. *IBM J. Res. Dev.*, 3(2):198 – 200, 1959. DOI: 10.1147/rd.32.0198 Cited on page(s) 21

[80] H.R. Lewis and Ch.H. Papadimitriou. *Elements of the Theory of Computation*. Prentice-Hall, 2nd edition, 1997. Cited on page(s) 23

[81] J. Friedl. *Mastering Regular Expressions*. O'Reilly Media, 3rd edition, 2002. Cited on page(s) 24

[82] Y. Bar-Hillel, M. A. Perles, and E. Shamir. On formal properties of simple phrase structure grammars. *Zeitschrift für Phonetik, Sprachwissenschaft und Kommunikationsforschung*, 14:143–172, 1961. Cited on page(s) 24

[83] M.O. Rabin. Probabilistic automata. *Information and Control*, 6:230–245, 1963. DOI: 10.1016/S0019-9958(63)90290-0 Cited on page(s) 24

[84] A. Kondacs and J. Watrous. On the power of quantum finite state automata. In *Proceedings of the 38th Annual Symposium on Foundations of Computer Science*, pages 66–75. IEEE Computer Society, Los Alamitos, 1997. DOI: 10.1109/SFCS.1997.646094 Cited on page(s) 26, 28, 30

[85] C. Moore and J. P. Crutchfield. Quantum automata and quantum grammars. *Theor. Comput. Sci.*, 237(1-2):275 – 306, 2000. DOI: 10.1016/S0304-3975(98)00191-1 Cited on page(s) 26, 27, 30

[86] A. Ambainis and R. Freivalds. 1-way quantum finite automata: strengths, weaknesses and generalizations. In *Proceedings of the 39th Annual Symposium on Foundations of Computer Science*, pages 332–341. IEEE, 1998. DOI: 10.1109/SFCS.1998.743469 Cited on page(s) 28

[87] A. Brodsky and N. Pippenger. Characterizations of 1-way quantum finite automata. *SIAM J. Comput.*, 31:1456–1478, 2002. DOI: 10.1137/S0097539799353443 Cited on page(s) 28

[88] A. Bertoni, C. Mereghetti, and B. Palano. Quantum computing: 1-way quantum automata. In *Developments in Language Theory*, volume 2710 of *LNCS*, page 162. Springer, Berlin/Heidelberg, Germany, 2003. DOI: 10.1007/3-540-45007-6_1 Cited on page(s) 30

[89] A. Ambainis, A. Ķikusts, and M. Valdats. On the class of languages recognizable by 1-way quantum finite automata. In A. Ferreira and H. Reichel, editors, *Proceedings of the 18th Annual Symposium on Theoretical Aspects of Computer Science*, volume 2010/2001 of *LNCS*, pages 75–86. Springer Berlin/Heidelberg, 2001. Cited on page(s) 30

[90] M. Ying. Automata theory based on quantum logic. (I). *Int. J. Theor. Phys.*, 39(4):985–995, 2000. DOI: 10.1023/A:1003642222321 Cited on page(s) 30

[91] M. Ying. Automata theory based on quantum logic II. *Int. J. Theor. Phys.*, 39(4):2545–2557, 2000. DOI: 10.1023/A:1026453524064 Cited on page(s) 30

[92] S. Gudder and R. Ball. Properties of quantum languages. *Int. J. Theor. Phys.*, 41:569–591, 2002. DOI: 10.1023/A:1015226708860 Cited on page(s) 30

[93] R. Lu and H. Zheng. Pumping lemma for quantum automata. *Int. J. Theor. Phys.*, 43(5):1191–1217, 2004. DOI: 10.1023/B:IJTP.0000048609.66662.87 Cited on page(s) 30

[94] J. Liu and Z.-W. Mo. Automata theory based on quantum logic: Recognizability and accessibility. *Int. J. Theor. Phys.*, 48(4):1150–1163, 2009. DOI: 10.1007/s10773-008-9888-6 Cited on page(s) 30

[95] A. Dawar. Quantum automata, machines and complexity. Slides available on-line at `http://www.cl.cam.ac.uk/~ad260/talks/warwick.pdf`, 2003. Cited on page(s) 30

[96] D. Qiu and L. Li. An overview of quantum computation models: Quantum automata. *Frontiers of Computer Science in China*, 2:193–207, 2008. DOI: 10.1007/s11704-008-0022-y Cited on page(s) 30

[97] M. Hirvensalo. Various aspects of finite quantum automata. In *Developments in Language Theory*, volume 5257 of *LNCS*, pages 21–33. Springer, Berlin/Heidelberg, Germany, 2008. DOI: 10.1007/978-3-540-85780-8_2 Cited on page(s) 30

[98] M. Golovkins. Quantum pushdown automata. In V. Hlaváč, K. Jeffery, and J. Wiedermann, editors, *SOFSEM 2000: Theory and Practice of Informatics*, volume 1963 of *LNCS*, pages 336–346. Springer Berlin / Heidelberg, 2000. DOI: 10.1007/3-540-44411-4 Cited on page(s) 30

[99] S. Gudder. Quantum computers. *Int. J. Theor. Phys.*, 39:2151–2177, 2000. DOI: 10.1023/A:1003698023288 Cited on page(s) 30

[100] M. Hirvensalo. *Quantum computing*. Springer-Verlag, Berlin, Germany, 2001. Cited on page(s) 33, 34, 38

[101] J. Riordan and C.E. Shannon. The number of two-terminal series-parallel networks. *Journal of Mathematics and Physics*, 21:83–93, 1942. Cited on page(s) 33

[102] O.B. Lupanov. A method of circuit synthesis. *Izvestia VUZ (Radiofizika)*, 1:120–140, 1958. Cited on page(s) 33

[103] U. Zwick. Scribe notes of the course boolean circuit complexity. Lecture notes available on-line at http://www.math.tau.ac.il/~zwick/scribe-boolean.html. Cited on page(s) 34

[104] T.F. Jordan. *Linear operators for quantum mechanics*. Dover Publications, 2006. Cited on page(s) 35

[105] E. Fredkin and T. Toffoli. Conservative logic. *Int. J. Theor. Phys.*, 21(3/4):219–253, 1982. DOI: 10.1007/BF01857727 Cited on page(s) 35

[106] C.H. Bennett. Logical reversibility of computation. *IBM J. Res. Dev.*, 17(6):525–532, 1973. DOI: 10.1147/rd.176.0525 Cited on page(s) 36, 43

[107] R. Landauer. Irreversibility and heat generation in the computing process. *IBM Journal of Research and Development*, 5(3):183–191, 1961. DOI: 10.1147/rd.53.0183 Cited on page(s) 36

[108] C.H. Bennett. Notes on Landauer's principle, reversible computation, and Maxwell's Demon. *Stud. Hist. Philos. Sci.*, 34(3):501 – 510, 2003. Cited on page(s) 36

[109] D. Deutsch. Quantum computational networks. *Proc. R. Soc. Lond. A*, 425:73, 1989. DOI: 10.1098/rspa.1989.0099 Cited on page(s) 36, 37, 66

[110] T. Toffoli. Bicontinuous extension of reversible combinatorial functions. *Math. Syst. Theory*, 14:13–23, 1981. DOI: 10.1007/BF01752388 Cited on page(s) 36

[111] A. Barenco, C. H. Bennett, R. Cleve, D. P. DiVincenzo, N. Margolus, P. Shor, T. Sleator, J. Smolin, and H. Weinfurter. Elementary gates for quantum computation. *Phys. Rev. A*, 52:3457, 1995. DOI: 10.1103/PhysRevA.52.3457 Cited on page(s) 37, 42, 43

[112] D. Deutsch, A. Barenco, and A. Ekert. Universality in quantum computation. *Proc. R. Soc. Lond.*, 449(1937):669–677, 1995. DOI: 10.1098/rspa.1995.0065 Cited on page(s) 42

[113] V. V. Shende, I. L. Markov, and S. S. Bullock. Minimal universal two-qubit controlled-NOT-based circuits. *Phys. Rev. A*, 69:062321, 2004. DOI: 10.1103/PhysRevA.69.062321 Cited on page(s) 42

[114] M. Möttönen, J. J. Vartiainen, V. Bergholm, and M. M. Salomaa. Quantum circuits for general multiqubit gates. *Phys. Rev. Lett.*, 93(13):130502, Sep 2004. DOI: 10.1103/PhysRevLett.93.130502 Cited on page(s) 42

[115] J. J. Vartiainen, M. Mottonen, and M. M. Salomaa. Efficient decomposition of quantum gates. *Phys. Rev. Lett.*, 92:177902, 2004. DOI: 10.1103/PhysRevLett.92.177902 Cited on page(s) 42

[116] Y. Lecerf. Machines de Turing réversibles. *Comptes rendus des séances de l'académie des sciences*, 257:2597–2600, 1963. Cited on page(s) 43

[117] B. Ömer. *Structured Quantum Programming*. PhD thesis, Vienna University of Technology, 2003. Cited on page(s) 45, 47, 48, 66, 68, 71, 74, 75

[118] D. Knuth. *The Art of Computer Programming*. Addison-Wesley, Reading, Massachusetts, 2nd edition, 1973. Cited on page(s) 45

[119] J. W. Backus, F. L. Bauer, J. Green, C. Katz, J. McCarthy, A. J. Perlis, H. Rutishauser, K. Samelson, B. Vauquois, J. H. Wegstein, A. van Wijngaarden, and M. Woodger. Revised report on the algorithmic language ALGOL 60. *Commun. ACM*, 6(1):1–17, 1963. DOI: 10.1145/366193.366201 Cited on page(s) 46

[120] E. Knill. Conventions for quantum pseudocode. Technical Report LAUR-96-2724, Los Alamos National Laboratory, 1996. Cited on page(s) 47, 48, 49, 66

[121] R. Cleve and D. P. DiVincenzo. Schumacher's quantum data compression as a quantum computation. *Phys. Rev. A*, 54(4):2636–2650, Oct 1996. DOI: 10.1103/PhysRevA.54.2636 Cited on page(s) 48

[122] T. H. Cormen, C. E. Leiserson, R. L. Rivest, and C. Stein. *Introduction to Algorithms*. The MIT Press, 2nd edition, 2001. Cited on page(s) 48

[123] R. Nagarajan, N. Papanikolaou, and D. Williams. Simulating and compiling code for the sequential quantum random access machine. *Electronic Notes in Theoretical Computer Science*, 170:101–124, 2007. DOI: 10.1016/j.entcs.2006.12.014 Cited on page(s) 51

[124] M. Elhoushi, M.W. El-Kharashi, and H. Elrefaei. Modeling a quantum processor using the QRAM model. In *Communications, Computers and Signal Processing (PacRim), 2011 IEEE Pacific Rim Conference on*, pages 409–415, 2011. DOI: 10.1109/PACRIM.2011.6032928 Cited on page(s) 51

[125] J. Lambek. How to program an infinite abacus. *Mathematical Bulletin*, 4(3):295–302, 1961. DOI: 10.4153/CMB-1961-032-6 Cited on page(s) 51

[126] M. Minsky. Recursive unsolvability of Post's problem of 'tag' and other topics in theory of Turing machines. *Annals of Mathematics*, 74(3):437–455, 1961. DOI: 10.2307/1970290 Cited on page(s) 51

[127] K. M. Svore, A. W. Cross, I. L. Chuang, A. V. Aho, and I. L. Markov. A layered software architecture for quantum computing design tools. *IEEE Computer*, 06(0018-9162):58–67, January 2006. DOI: 10.1109/MC.2006.4 Cited on page(s) 53, 54, 55, 56, 57, 66

[128] K. M. Svore, A. W. Cross, A. V. Aho, I. L. Chuang, and I. L. Markov. Toward a software architecture for quantum computing design tools. In P. Selinger, editor, *Proceedings of the 2nd International Workshop on Quantum Programming Languages*, 2004. Cited on page(s) 54

[129] S. Bettelli. *Toward an architecture for quantum programming*. PhD thesis, Università di Trento, February 2002. Cited on page(s) 55, 68

[130] S. Bettelli, L. Serafini, and T. Calarco. Toward an architecture for quantum programming. *Eur. Phys. J. D*, 25(2):181–200, 2003. DOI: 10.1140/epjd/e2003-00242-2 Cited on page(s) 55, 66, 67, 68, 70

[131] I. L. Chuang. Quantum circuit viewer: qasm2circ. Software available from the web page http://www.media.mit.edu/quanta/qasm2circ/. Cited on page(s) 55, 56, 57

[132] T. Altenkirch and J. Grattage. A functional quantum programming language. In *Proceedings. 20th Annual IEEE Symposium on Logic in Computer Science*, pages 249–258. IEEE, 2005. DOI: 10.1109/LICS.2005.1 Cited on page(s) 59, 89

[133] J. Grattage. An overview of QML with a concrete implementation in Haskell. *Electronic Notes in Theoretical Computer Science*, 270(1):157–165, 2011. Proceedings of the Joint 5th International Workshop on Quantum Physics and Logic and 4th Workshop on Developments in Computational Models (QPL/DCM 2008). DOI: 10.1016/j.entcs.2011.01.015 Cited on page(s) 59, 89, 90, 92

[134] H. Rosé, T. Asselmeyer-Maluga, M. Kolbe, F. Niehoerster, and A. Schramm. The fraunhofer quantum computing portal - www.qc.fraunhofer.de - a web-based simulator of quantum computing processes. 2007. Cited on page(s) 59

[135] J. A. Miszczak and P. Wycisk. Software available from the web page http://zksi.iitis.pl/wiki/software:qml. Cited on page(s) 60

[136] B. Ömer. A procedural formalism for quantum computing. Master's thesis, Vienna University of Technology, 1998. Cited on page(s) 62, 68, 70, 71, 85

[137] B. Ömer. Quantum programming in QCL. Master's thesis, Vienna University of Technology, 2000. Cited on page(s) 62, 68, 71, 85

[138] S. Balensiefer, L. Kregor-Stickles, and M. Oskin. An evaluation framework and instruction set architecture for ion-trap based quantum micro-architectures. In *Proceedings of the 32nd annual international symposium on Computer Architecture*, ISCA '05, pages 186–196, Washington, DC, USA, 2005. IEEE Computer Society. DOI: 10.1145/1080695.1069986 Cited on page(s) 63

[139] R. van Meter. State of the art in quantum computer architectures, 2011. Document available from http://aqua.sfc.wide.ad.jp/publications/van-meter-quantum-architecture-handout.pdf. Cited on page(s) 63

[140] N.C. Jones, R. van Meter, A.G. Fowler, P.L. McMahon, J. Kim, T. D. Ladd, and Y. Yamamoto. Layered architecture for quantum computing. *Arxiv preprint arXiv:1010.5022*, 2010. Cited on page(s) 63

[141] D.P. DiVincenzo and D. Loss. Quantum information is physical. *Superlattices and Microstructures 23*,, 23:419, 1998. DOI: 10.1006/spmi.1997.0520 Cited on page(s) 65

[142] L. K. Grover. Quantum computers can search rapidly by using almost any transformation. *Phys. Rev. Lett.*, 80:4329–4332, 1998. DOI: 10.1103/PhysRevLett.80.4329 Cited on page(s) 66

[143] M. Mosca. *Quantum Computer Algorithms*. PhD thesis, Wolfson College, University of Oxford, 1999. Cited on page(s) 66

[144] C. Bennett and S.J. Wiesner. Communication via one- and two-particle operators on Einstein-Podolsky-Rosen states. *Phys. Rev. Lett.*, 69:2881–2884, 1992. DOI: 10.1103/PhysRevLett.69.2881 Cited on page(s) 66

[145] G. Brassard, A. Broadbent, and A Tapp. Quantum pseudo-telepathy. *Found. Phys.*, 35:1877–1907, 2005. DOI: 10.1007/s10701-005-7353-4 Cited on page(s) 66

[146] D. Bacon and W. van Dam. Recent progress in quantum algorithms. *Commun. ACM*, 53(2):84–93, 2010. DOI: 10.1145/1646353.1646375 Cited on page(s) 66

[147] E. H. Knill and M. A. Nielsen. *Encyclopedia of Mathematics, Supplement III*, chapter Theory of quantum computation. Kluwer, 2002. Cited on page(s) 66

[148] W. Mauerer. Semantics and simulation of communication in quantum programming. Master's thesis, University Erlangen-Nuremberg, 2005. Cited on page(s) 66, 69, 70, 85, 89

[149] H. Mlnařík. Operational semantics and type soundness of quantum programming language LanQ. Technical report, Masaryk University, 2009. Cited on page(s) 66, 71

[150] H. Mlnařík. Semantics of quantum programming language LanQ. *Int. J. Quant. Inf.*, 6(1, Supp.):733–738, 2008. DOI: 10.1142/S0219749908004031 Cited on page(s) 66, 68, 71

[151] S. Gay. Quantum programming languages: Survey and bibliography. *Bulletin of the European Association for Theoretical Computer Science*, 2005. DOI: 10.1017/S0960129506005378 Cited on page(s) 66, 70

[152] D. Unruh. Quantum programming languages. *Informatik – Forschung und Entwicklung*, 21(1–2):55–63, 2006. DOI: 10.1007/s00450-006-0012-y Cited on page(s) 66, 70

[153] R. Rüdiger. Quantum programming languages: An introductory overview. *Comput. J*, 50(2):134–150, 2007. DOI: 10.1093/comjnl/bxl057 Cited on page(s) 66, 70

[154] H. Weimer. The C library for quantum computing and quantum simulation. Software avaible from the web page http://www.libquantum.de/, 2003–. Cited on page(s) 68

[155] H. Mlnařík. *Quantum Programming Language LanQ*. PhD thesis, Masaryk University, 2007. Cited on page(s) 68, 70, 71, 77, 80

[156] P. Maymin. Extending the lambda calculus to express randomized and quantumized algorithms. 2008. Cited on page(s) 69

[157] A. van Tonder. A lambda calculus for quantum computation. *SIAM J. Comput.*, 33(5):1109–1135, 2004. DOI: 10.1137/S0097539703432165 Cited on page(s) 69, 83, 84

[158] J. Karczmarczuk. Structure and interpretation of quantum mechanics: a functional framework. In *Proceedings of the ACM SIGPLAN workshop on Haskell*, pages 50–61. ACM Press, 2003. DOI: 10.1145/871895.871901 Cited on page(s) 69, 85

[159] P. Selinger. Towards a quantum programming language. *Math. Struct. Comp. Sci.*, 14(4):527–586, 2004. DOI: 10.1017/S0960129504004256 Cited on page(s) 69, 70, 85

[160] J. Grattage. *QML: A functional quantum programming language*. PhD thesis, School of Computer Science and School of Mathematical Sciences, The University of Nottingham, 2006. Cited on page(s) 70, 89

[161] D.A. Sofge. A survey of quantum programming languages: History, methods, and tools. In D. Avis, C. Kollmitzer, and V. Ovchinnikov, editors, *Second International Conference on Quantum, Nano and Micro Technologies, 2008*, pages 66–71, 2008. Cited on page(s) 70

[162] List of QC simulators. Web page available at http://www.quantiki.org/wiki/List_of_QC_simulators, 2005–. Cited on page(s) 70, 85, 92

[163] S. Abramsky and B. Coecke. A categorical semantics of quantum protocols. In *Proceedings of the 19th IEEE conference on Logic in Computer Science (LiCS'04)*. IEEE Computer Science Press, 2004. DOI: 10.1109/LICS.2004.1 Cited on page(s) 70

[164] S. Abramsky and B. Coecke. *Handbook of Quantum Logic and Quantum Structures*, volume II, chapter Categorical quantum mechanics. Elsevier, 2008. Cited on page(s) 70

[165] B. Coecke. Quantum picturalism. *Contemporary Physics*, 51:59–83, 2010. DOI: 10.1080/00107510903257624 Cited on page(s) 70

[166] P. Selinger. Dagger compact closed categories and completely positive maps. In *Proceedings of the 3rd International Workshop on Quantum Programming Languages*, 2005. Chicago, June 30–July 1 (2005). DOI: 10.1016/j.entcs.2006.12.018 Cited on page(s) 70

[167] B. Ömer. QCL – a programming language for quantum computers. Software available on-line at `http://tph.tuwien.ac.at/~oemer/qcl.html`. Cited on page(s) 71, 72

[168] H. Mlnařík. LanQ – a quantum imperative programming language. Software available on-line at `http://lanq.sourceforge.net`. Cited on page(s) 71, 77

[169] I. Glendinning and B. Ömer. Parallelization of the QC-lib quantum computer simulator library. In R. Wyrzykowski, J. Dongarra, M. Paprzycki, and J. Wasniewski, editors, *Parallel Processing and Applied Mathematics*, volume 3019 of *Lecture Notes in Computer Science*, pages 461–468. Springer, 2004. DOI: 10.1007/b97218 Cited on page(s) 72

[170] D. Deutsch and R. Jozsa. Rapid solution of problems by quantum computation. *Proc Roy Soc Lond A*, 439:553–558, 1992. DOI: 10.1098/rspa.1992.0167 Cited on page(s) 75

[171] W. K. Wootters and W. H. Zurek. A single quantum cannot be cloned. *Nature*, 299:802–803, 1982. DOI: 10.1038/299802a0 Cited on page(s) 80, 86

[172] V. Vedral, A. Barenco, and A. Ekert. Quantum networks for elementary arithmetic operations. *Phys. Rev. A*, 54:147–153, 1996. DOI: 10.1103/PhysRevA.54.147 Cited on page(s) 80

[173] J. Hughes. Why functional programming matters? *Comput. J*, 32(2):98–107, 1989. DOI: 10.1093/comjnl/32.2.98 Cited on page(s) 83

[174] P. Selinger. A brief survey of quantum programming languages. In *Proceedings of the 7th International Symposium on Functional and Logic Programming*, volume 2998 of *LNCS*, pages 1–6, 2004. DOI: 10.1007/978-3-540-24754-8_1 Cited on page(s) 83

[175] K. Hinsen. The promises of functional programming. *Comput. Sci. Eng.*, 11(4):86–90, 2009. DOI: 10.1109/MCSE.2009.129 Cited on page(s) 84

[176] Shin-Cheng Mu and Richard Bird. Functional quantum programming. In *Asian Workshop on Programming Languages and Systems*, KAIST, Dajeaon, Korea, dec 2001. Cited on page(s) 84

[177] J. Skibiński and H. Thielemann. Numeric quest. Software available on-line at `http://www.haskell.org/haskellwiki/Numeric_Quest`. Cited on page(s) 84

[178] A. van Tonder and M. Dorca. Quantum computation, categorical semantics and linear logic. Preprint arXiv:quant-ph/0312174. Cited on page(s) 84

[179] J.Y. Girard. Linear logic. *Theor. Comput. Sci.*, 50(1):1–101, 1987. DOI: 10.1016/0304-3975(87)90045-4 Cited on page(s) 84

[180] A. van Tonder. A lambda calculus for quantum computation. Software available on-line `http://www.het.brown.edu/people/andre/qlambda/`. Cited on page(s) 84

[181] A. Sabry. Modeling quantum computing in Haskell. In *ACM SIGPLAN Haskell Workshop*, 2003. DOI: 10.1145/871895.871900 Cited on page(s) 85

[182] J. Grattage. QML@CS.Nott. Software available from the web page `http://sneezy.cs.nott.ac.uk/QML/compiler/`. Cited on page(s) 89

[183] The Glasow Haskell Compiler, 1989-. Software available from the web page `http://www.haskell.org/ghc/`. Cited on page(s) 89

[184] J. Karczmarczuk. Scientific computation and functional programming. *Comput. Sci. Eng.*, 1(3):64–72, 1999. DOI: 10.1109/5992.764217 Cited on page(s) 92

[185] P. Nyman. A symbolic classical computer language for simulation of quantum algorithms. In P. Bruza, D. Sofge, W. Lawless, K. van Rijsbergen, and M. Klusch, editors, *Quantum Interaction*, volume 5494 of *LNCS*, pages 158–173. Springer Berlin/Heidelberg, 2009. DOI: 10.1007/978-3-642-00834-4 Cited on page(s) 92

[186] A. Kissinger, A. Merry, B. Frot, B. Coecke, L. Dixon, M. Soloviev, and R. Duncan. Quantomatic. Software available on-line at `http://sites.google.com/site/quantomatic/`. Cited on page(s) 92

[187] L. Dixon and R. Duncan. Graphical reasoning in compact closed categories for quantum computation. *Ann. Math. Artif. Intell.*, 56:23–42, 2009. DOI: 10.1007/s10472-009-9141-x Cited on page(s) 92

[188] J. McCarthy. Recursive functions of symbolic expressions and their computation by machine, Part I. *Commun. ACM*, 3(4), 1960. DOI: 10.1145/367177.367199 Cited on page(s) 93

Author's Biography

JAROSŁAW ADAM MISZCZAK

Dr. Jarosław Adam Miszczak is a researcher at the Institute of Theoretical and Applied Informatics of the Polish Academy of Sciences in Gliwice, Poland. He obtained his Master's degree with specialization in theoretical physics from University of Silesia in Katowice, Poland, in 2005 and Ph.D. degree in Computer Science from the Institute of Theoretical and Applied Informatics of the Polish Academy of Sciences, Gliwice, Poland, in July 2008. His research interests include quantum information theory, foundations of quantum mechanics, scientific computing, and theory of programming languages.